PRAISE FOR NEGOTIATE WISELY IN BUSINESS & TECHNOLOGY

"An action-packed read with lots of real-life examples. This building block approach to business negotiations guides us through all the practical techniques we need that improve our results."

Michael Meriton, President and CEO,
Financial Technologies International, Inc.

"Mladen and Harvey are master practitioners of negotiations. They have succeeded in creating a practical primer for leaders, negotiators and sales people on one of the most important topics for anyone in business."

Keith Blackwell, Founder, Chairman and CEO,
Bristol Technology, Inc. (now part of HP)

"Negotiation is a discipline that can be learned and practiced. Blending personal traits and psychological factors involved in negotiations with business knowledge makes *Negotiate Wisely* a compelling and seminal work on the subject."

Richard J. Kaplan, Vice President and Assistant
Vice President and General Counsel, IBM Global Services

"K&R has put the science to the art of negotiation. A must read for companies and individuals who wish to improve their results in specific opportunities and more importantly, those who look for skills that help them shape enduring and profitable business relationships."

Patrick Hennessy
President, Vertical Markets, Merkle Inc.

"Kresic and Rosen created a joy to read book from beginning to end. As you progress, you realize you have in your hands the genuine Bible of effective and successful negotiations. You will learn a lot from *Negotiate Wisely*."

Tom McClain, Retired CEO E-Commerce
And I.T., Spencer Trask,
Executive VP, Dun & Bradstreet

NEGOTIATE WISELY
IN BUSINESS & TECHNOLOGY

Includes new and updated information
from the original work.

MLADEN D. KRESIC

Based on original work co-authored with Harvey I. Rosen.

Foreword by John Patrick

Fusion Marketing Press
Colorado Springs, CO

ISBN-13:978-0-9968789-0-6

Fusion Marketing Press
Colorado Springs, CO

DEDICATION

To my wife Maryjane and my children, Briana, Will, Alex and Meredith, your devotion and support are the foundations I stand on. I love and thank you.

Original Dedication (2004)

To our beloved wives, Alice Rosen and Maryjane Kresic, whose patience and support gave us the freedom to create.

ACKNOWLEDGEMENTS

The current version of this book has evolved over the past three years. I owe a debt of gratitude to my colleague and friend Gerard Francois for his enthusiasm and dedication in that effort. Thanks to all of our K&R team members as well as our colleagues from Fusion Marketing, including Gail Carson whose editing and advice have been invaluable.

Original Acknowledgement (2004)

This book was long in the making. It had its formative stages when K&R began in 1994. After many fits and starts we have *Negotiate Wisely*. Thank you to the following: The K&R team, especially to Tim Delaney with his insights and Susi Petersen who kept the process moving; to Laurie Rozakis, a wonderful editor with great humor and skill with the word. And of course, Peter Faulkner, thanks for your constant commentary to keep us going, to get us to print and out to our audience.

CONTENTS

BIOGRAPHIES

Mladen Kresic is President and CEO of K&R Negotiation Associates LLC. He has spent most of his career negotiating business relationships in the technology sector. He represents clients throughout the world in all types of negotiations including sales, licensing, strategic alliances, and mergers. Mladen is a former counsel for the IBM software business. He is the author of numerous articles on business and negotiations published online and in journals like Thompson Reuters, the Hartford Business Journal and Forbes.com. He has also been a lecturer in over 40 countries around the world and a guest speaker at institutions such as the MIT Enterprise Forum, the Yale School of Business and the International Licensing Executives Society. Mladen also served as general counsel to Bristol Technology prior to its acquisition by Hewlett Packard and on the boards of several technology companies, including Bristol.

As a technology lawyer, Mladen is a member of the New York, California, and Connecticut bars. He lives with his wife in Puerto Rico, often visiting his grown children in Connecticut and New York.

IN MEMORIAM

Harvey Rosen was co-founder (with Mladen) of K&R Negotiation Associates LLC. Prior to his premature passing in 2007, Harvey was considered one of the premier international technology negotiators. After a successful sales and sales management career with IBM, he became the company's "go-to" lead negotiator for OEM transactions that included hardware, software, services and support. Following his retirement from IBM, Harvey continued to represent IBM and other companies as a lead negotiator.

As a competitive athlete, Harvey viewed teamwork in negotiations as a key element to their success. As a testament to his skillful team approach, his transactions usually stood the test of time, often lasting ten or more years and expanding in scope.

Harvey is survived by his three children, his grandchildren, and his wife, Alice.

ABOUT THIS BOOK

We first published *Negotiate Wisely* in 2004 after years of coaching and participating in business negotiations. We felt it was time to re-release this book to reinforce what works and to add a few updates as we have continued to learn and expand. Before and since 2004 we had written a wealth of material in articles and workbooks used to provide advice and negotiation training to our clients. This book is a result of pulling together all the work and expanding it into a practical guide to the discipline of business negotiations. It is written using a building block approach. First, we build a solid base of what you need to know. Then, we expand the base through practical examples and deal forensics. Finally, we add chapters on tools, techniques, tactics and teamwork designed to provide you with the means to obtain the transaction results and relationships you desire.

The design of this book should encourage you to use it as a negotiation reference guide. For example, if you are in the middle of your negotiation and you think the other side is "gaming" you, go to Chapter 13 on tactics. Or, if your internal teamwork is not working the way you think it should, go to Chapter 15 on teamwork, and get the reminders you need. Most importantly, if you are having difficulties creating your initial offer, go to Chapter 7. And, if you are having trouble articulating the value for the offer you have made, refer to Chapter 4 before you continue.

Negotiate Wisely will see you through your most challenging business opportunities. To that end, we hope that you will come back to it time and time again.

Mladen D. Kresic

FOREWORD

Negotiations are pervasive in both our business and personal lives. Whether selling products, making a deal to sell a business, starting a new business, launching a new product, or renting a tent for your daughter's wedding to be held at your home, we find ourselves negotiating. Even though we may not consciously think of it, negotiations are indeed part of our daily lives.

In today's business and technological world, negotiations can run the gamut from simple to complex and from friendly to nasty. You can love it or loathe it, but we all negotiate in some manner, even though we may not have the skills to do it well. As with accounting, marketing, swimming or horseback riding, negotiating requires significant skills. Developing and honing those skills is essential to enable your negotiations to be as effective as possible. That is a matter of attitude. Then your ability to negotiate will become a powerful asset to assist you in achieving your goals.

Given how much of our life is spent negotiating, we should enjoy the process. And you can enjoy the process more if you know what you are doing, what you are after, and if your negotiations result in good relationships. I was recently reminded of this in a personal negotiation. Shortly after the beginning of year 2000, I began negotiating with a local builder to construct a new home. I wanted to buy a piece of land, hire an architect to design the home, and hire the builder to build it. The builder wanted to own the land because he wanted to control all aspects of the project. He also wanted to establish the design from one that he had previously built. Of course, he also wanted to establish the price to be paid for the home. We could not have had more polar positions at the outset.

But what did the builder really have to have? And what was I worried about? After a number of meetings, it became clear that we would not be able to agree on a price up front. The builder was concerned that he would not make money. I was concerned that I would not get exactly the house I wanted. And costs could run out of control. The builder only builds a couple of houses per year and although he wanted to build my house, he was clearly not dependent on me for his livelihood. How do I get control over the design while he has control over the construction of the project?

Throughout the "discussions", the builder and I got to know each other quite well and we established a high level of mutual trust. How to negotiate the control? In thinking through the goals we each had and openly discussing them, we realized that these goals were not in conflict. This led to a win-win arrangement. The architect would be "our" architect. He would establish the design but would not establish the specific materials, processes or techniques. Those would be up to the builder but we would confer on critical items. The price issue was moved off the table by agreeing on a general contractor fee. Each month I would pay a pro rata share of the fee plus the materials and labor that had been expended the prior month. The contractor fee assured the builder that he would not lose money and the time and material approach assured me that I would get what I wanted. We both had the controls we wanted. In the end, we are delighted with the house and the builder has become a good personal friend. Taking the time to discuss and understand each other's goals turned conflict into a great deal and a great relationship.

Negotiating skills are critical for success in your personal and business life. It is not just about winning a deal. In fact, if you think of winning as the only criteria, you may not gain all that you can. Comprehensive and well thought out negotiations cannot only win the deal but also lead to improved business results, a sustainable competitive advantage, new relationships, and new efficiencies in development, manufacturing or distribution. Effective negotiating must encompass specific objectives, and by incorporating broad thinking into your negotiations, you can also get new products to market and achieve their financial goals, assure continued availability of capital for growth and innovation, create new promotional opportunities, have a positive impact on morale, and establish long term working relationships.

Despite the critical role negotiations play in business and life, many people have not taken the time to hone their skills. Some may feel that negotiating well is an art or perhaps something one is born with. Others may believe that the key to negotiating is to be able to talk louder or more aggressively or to be able to maintain a "poker face", These types of actions may influence negotiations; but if you want to systematically and consistently maximize the full potential of negotiating, it is essential to learn the fundamentals and beyond. In fact, if you are already a good negotiator, wouldn't it be great to know what makes you good, so you can systematically repeat your negotiation success or be an effective mentor to others?

The best way to learn a practical approach to negotiations is from the experts. That is who Harvey and Mladen are. They have merged more than fifty years of experience into a definitive work called *Negotiate Wisely*. The book takes the emotion out of the process and systematically explains how to be an effective negotiator. Numerous real-world examples show you how to deal with an adversarial negotiator. The tutorial approach of *Negotiate Wisely* can help you develop professional techniques to shorten negotiation cycles and build solid relationships while achieving your specific business or personal goals.

The best part is that *Negotiate Wisely* helps you adapt your style to maximize achievement of your goals—without changing who you are.

John Patrick

Retired IBM Vice President and Chief Internet Technology Officer, founding member of the World Wide Web Consortium at MIT. Founding Member and Chairman of the Global Internet Project. He is also considered "Father of the ThinkPad" (Wired News, Computerworld) and the leader of IBM's e-metamorphosis (Harvard Business Review).

Get comfortable.

Stretch out and relax. Grab your beverage of choice and settle into an easy chair. We are going to get right into the nuts and bolts of negotiation.

Your curiosity is about to get piqued. You have a major negotiation on your hands. You want to be thoughtful, do a good job and have fun at the same time.

> *You are the chief negotiator for the Catco Corporation, a pest eradication company. Your soon-to-be-released Intelligent Eradication Systems (IES) technology has been touted as revolutionary by the industry. That's great, because you are in a race with two other companies that are also developing potentially effective technologies.*

> *Right now, your company lacks manufacturing and marketing resources, but if IES becomes the new standard, you should have no trouble licensing it for a great deal of money. You would also find it easy to pay ambitious companies to manufacture the product for you.*

> *During the last few weeks, you have been negotiating a huge deal with Infest City, which has a major problem with rodents. A new strain of three-thumbed mice has apparently developed immunities to all existing pest-eradicating technologies, and the little rodents are overrunning the town. The mayor, Julianna Rooda, wants these pests under control by the time she is up for reelection fifteen months from now. That means she wants the Trap Solution developed now!*

> *To reach that goal, Mayor Rooda has contacted Catco and one of its competitors, Ratco. Inc., the second largest company in the pest eradication business. Ratco has a huge sales and service team and ample manufacturing capability. It has a well-established and loyal customer base. However, Ratco's technology is no longer considered*

leading edge. Everyone knows that Ratco's developers don't know how to deal with three-thumbed mice. Ratco's CEO would like to acquire an up-and-coming technology-oriented company; but that does not look too promising as Ratco's stock is undervalued. In addition, the company's cash reserves are pretty low. In an attempt to increase consumer confidence, the CEO made public statements that the company will definitely be offering premium products in the near future.

Mayor Rooda insists that Catco and Ratco work together on the Trap Project, the project designed to rid Infest City of the three-thumbed mice. Last month, Mayor Rooda met privately with representatives of each company. She rebuffed attempts by both sides to work with other companies, although neither side accepted this decision as final. When it became clear that they were probably stuck with each other if they wanted to work with the city, Catco and Ratco agreed to create a new project-specific entity, which they are calling NewCo. You have already agreed that Catco and Ratco will each own 50% of NewCo. Obviously, Catco will contribute its IES technology, and Ratco will do the manufacturing and most of the marketing.

Ratco has further agreed—in general principle—to manufacture some other pest control products for you. You have agreed —also in vague principle—to license some of your technologies to Ratco for uses that would provide Catco with incremental revenues. Now you are trying to iron out the details, and the meetings have not gone well. Raised voices and finger pointing have become the norm.

The main points of contention are:

1. Who gets rights to any technologies developed as a result of the Trap Project?

2. Who distributes products to Infest City residents? Who services these products?

3. Who may distribute what products and service them outside Infest City?

4. *Who gets to appoint the president and development manager of NewCo?*

5. *When will the staff of NewCo come from? Who pays the bill? Most importantly...*

6. *What is the revenue flow? How are the profits divided?*

Obviously, you need the numbers to get specific about these answers. But let's ignore the specifics at this time and look at the big picture. Ideally, you would like to find a win-win situation. In fact, you will have to do that if you want the deal to work because neither Catco nor Ratco are in the business of losing.

But what does winning mean in this situation? Clearly, there are many ways for both companies, and the city, to come out winners. Win-win doesn't necessarily mean equal gains for all sides, but it does mean positive outcomes for them. You would like Catco to be the winner with a big "W". And if Ratco and the city also feel that they got a big "W", it would be a bonus. Naturally, Ratco's hardball negotiation team has a different idea. Their history tells us that they would like Ratco to be the winner with a big "W":

Ratco: WIN Catco: win

How would you approach this situation? You are going to consider questions such as:

- What *MUST* Catco get out of this deal?

- From Catco's point of view, what must Ratco *NOT* get out of this deal?

If you are going to be a successful negotiator, you will have to anticipate the other side's positions. For example;

- What *MUST* Ratco get out of this deal?

- From Ratco's point of view, what must Catco *NOT* get out of this deal?

Good news! In this book, we'll cover methods that will help you develop a logical approach to this type of negotiation dilemma and enable you to get closer to closing a deal that works for you. This will occur as you learn the **K&R Negotiation Method™**[1] throughout this book. The K&R Negotiation Method is our process for preparation, planning and execution of negotiations. You will discover that there are different and probably more effective ways to evolve your own position and your best estimate of the other side's position. As you go through the book, don't be surprised if you find yourself reviewing steps you have taken in your own negotiations and saying, "Now, why didn't I think of that?"

We promise you that by the time you reach the last page of this book, you will think of that. The "that" will be the K&R Negotiation Method™ tools you need to have a successful negotiation.

This book will teach you how to be a more effective negotiator. We cover all the bases, providing examples drawn from our collective years of experience at the negotiation table. We provide you with the tools that take much of the guesswork out of negotiating and replace it with logic. If you use the K&R Negotiation Method, your negotiating skills will improve, your confidence will increase and your chances of negotiation success will rise.

Before we begin, let's get a few general points straight. The journey this book takes us through is a logical building block approach. Yet in real life, negotiations are an iterative process. For example, you don't prepare only before you meet with the customer; you continually prepare and gather new data throughout the process to refine your strategy. You don't evaluate **K&R's Negotiation Success Range (NSR)™**—explained in Chapters 1 and 7 - once, but continue to refine it throughout the process.

This is a practical book written from experience. It is not theory-spouting

---

[1] The K&R Negotiation Method is part of the overall K&R Win Wisely Method™ for sales engagement.

rules. Because negotiations are inherently unique, involving a lot of human factors, there are exceptions to every general rule we give you. That's why one of our favorite phrases is "it depends". It also explains why we rarely use the words "always" or "never". In the end, it's your judgment in each situation that will rule the day. What we hope to do is sharpen the way you make those judgments. We give you some systematic methods and tools with which you can improve your portfolio of pertinent information to make sound, informed and conscious decisions.

As we suggest in many places, much of what we talk about is rooted in human psychology—predictability of how people are likely to behave in given situations. That does not mean that we are certain they will always behave that way. So, again, your informed judgment is key.

We also tell you that there is no exact way to behave as a negotiator. We say, "Be yourself. It's who you do best." That doesn't mean there are no general guidelines you should follow. Being yourself refers to personality traits that make your style unique. If you are quiet, for example, it's difficult to all of a sudden be boisterous. If you are not funny, you probably can't suddenly become funny. That's okay, we know excellent negotiators with all different styles.

Use your particular style, your particular strength. Play to your strengths. For example, if you are a morning person, don't schedule the important calls or meetings late in the day. Likewise, if you are an evening person or night owl, try not to make your negotiations for early in the day.

Throughout this book, you will see us use the concept of repetition in a number of ways. We believe repetition is essential to effective learning as well as effective persuasion in negotiations. Interestingly, you'll also see how repetition is effective in gathering information.

People often ask us why certain deals succeed and others fail. As with many aspects of negotiations, it depends. In this book, we will review both deals that succeeded and ones that didn't. We'll take you through many actual negotiation situations. We'll review the results, both good and bad. Then we'll summarize what worked and what didn't. This approach will help you understand many of the important negotiation principles in context. You will then be able to apply these practical approaches in your negotiations, and improve your results.

While this book is written with many examples for business negotiators, the general concepts apply to all negotiations, personal and professional. How you choose to apply them is up to you! Lastly, though you may not want the other side to read this book, you'll have fun reading it yourself!

Let's take a sneak peek to see what we'll cover.

In *Chapter 1*, you'll be reminded that in order to negotiate a deal that works best, you must understand your own needs and the needs of the people on the other side. You'll explore some roadblocks to success and the effects of emotions, patience and teamwork on successful negotiations. We'll also help you take a close look at your own business skills. We think you'll find that you have far more strengths as a negotiator than you realize! Finally, you'll learn how to draw on these strengths and how to compensate for your weaknesses.

In *Chapter 2*, you'll explore the question: Is negotiation an art, a science or a combination of the two? We'll introduce you to the key concepts of **principled concessions** and **quantifiable value**. Then we give you an overview of the K&R Negotiation Method™.

Chapter 3 is devoted to the issue of **effective communication**, one of the cornerstones of all successful negotiations. You'll learn about the different types of communication and how to communicate responsibly. We'll teach you K&R's version of **P&L—patience** and **listening**. Patience and listening are two keys to success. We'll also explore the effects of poor listening habits on your **credibility**.

In *Chapter 4*, you'll build on what you learned about credibility in Chapter 3 to explore its relationship to **leverage**. You'll discover what leverage is, and what it is not. We'll also teach you how to use leverage to help you make a **value argument** and build a case.

Chapter 5 reinforces the lessons of Chapters 3 and 4 with four scenarios that hinge on credibility and leverage. In each of these real-life negotiating examples, you'll get the chance to apply your skills and your learning thus far about the K&R Negotiation Method™. In the strategy session at the end of the chapter, you can compare your resolutions to some of ours. The chapter concludes with a discussion of the importance of **teamwork**. This is another central element in our methodology.

In *Chapter 6*, you'll learn about **K&R's Six Principles™** to successful nego-tiations. They're so important that we'll give you a sneak peek here:

1. Get M.O.R.E.—Preparation is key to a winning negotiation.

2. Protect your weaknesses; utilize theirs.

3. A team divided is a costly team.

4. Concessions easily given appear of little value.

5. Negotiation is a continuous process.

6. Terms cost money; someone pays the bill.

Chapter 7 explains **K&R's Negotiation Success Range (NSR)™**—one of our most useful tools. You'll learn how to use this indispensable tool to figure out the price and terms range that help you make credible offers in a negotiation. And you'll learn about principled concessions and how they can help you and your counterparts develop confidence and accelerate the process to closure.

In *Chapter 8* we'll review how to gather the information you need to negotiate smarter. We'll cover what kinds of information are valuable and how to evaluate credible sources of information. The chapter concludes with a discussion of going from merely communicating to being believable and persuasive. We'll show you how to understand customer motivations and objectives. Building value is your way of getting a customer to come to the decision you want. Build value and you build your case.

Chapter 9 discusses how nearly everything we do in a negotiation affects either **credibility** or **leverage**—and usually both. We'll teach you how to manage, block and utilize information.

Chapter 10 focuses on **K&R's MID™,** a must-have technique for prioritizing issues in negotiations that helps us also distinguish between **means** and **ends**. We'll explore the means—alternate ways of accomplishing **goals**— and the ends, the goals themselves. We'll show you how to tell the difference among mandatory, important and desirable goals. The MID analysis forces you to figure out what problem you are really trying to solve. It can also help you sort through the bargaining in a negotiation so you can make meaningful tradeoffs to close deals sooner with better results. To truly gain the advantages of the MID, you must listen to gain understanding, not to argue. Remember Catco, Ratco and Infest City? We'll also get back to that scenario in this chapter.

Chapter 11 discusses how effective persuasive communication contains company value and personal value. You'll learn how to bargain and make concessions using logical reasoning, stimulus/response, repetition, positive inducements, negative inducements and tactics. Finally, we'll show you how to understand business goals and analyze the selling and environment to enhance your negotiation posture.

Chapter 12 covers the negotiator's responsibilities including understanding and communicating management goals, establishing process, and establishing support roles and rules. We'll teach you how to earn trust with your team, including management, as well as how to address and manage internal conflict. You'll also explore a phenomenon we call **Momentum to Closure**.

Chapter 13 explores different types of tactics, including stimulus/response, the "chess match", the psychological stake, and the power of "face". We'll help you decide if using humor to relieve pressure will work for you. We'll conclude with a discussion of managing tactics and gamesmanship.

In *Chapter 14*, you'll learn all about internal and external negotiations. For internal negotiations, we cover cross-functional motivations; for external negotiations, we discuss customer-functional motivations. You'll get a chance to practice what you learn through two real-life negotiation scenarios. We also cover the concept of managing the macro and micro agendas in negotiations. That means we'll learn how to manage the overall negotiation process as well as a single meeting or phone call with the other side—all with a goal of closing faster with better results.

Chapter 15 covers teamwork, the best "tactic". We'll explain how to recognize varied motivations, both internally and externally. You'll discover why it's important to debate internally, but essential to unify externally. Then comes a discussion on building two-way trust and responsibility with management and creating a true team in which everyone contributes expertise. This chapter also includes information on working as a team and communicating openly. Last, you'll learn why some deals succeed...but others flop.

USING THE COMPANION WORKBOOK

In our decades of training negotiation professionals, we have found that stopping occasionally to think through and apply what you have read enhances the "stickiness" of the subject material and helps grow skills faster.

Throughout the book, you will find many references to the companion workbook. Like all of the content in this book, which of these exercises you use—and in what order—is completely up to you! However, we highly recommend working through the sample scenarios whenever you can to increase your comfort with the concepts and deepen your understanding of the negotiation examples presented here. The process is simple: If you choose to work through the exercises, stop reading when you see the icon below, and open the workbook. If you read ahead before completing the exercise, you will quickly encounter the answers. Once you complete each exercise, refer back to the chapter for the answers, the K&R Deal Forensics and further discussion.

CHAPTER 1: ABOUT YOU AND YOUR VALUE

So, let's talk about you, shall we?

Believe it or not, we have a pretty good idea about who you are. Regardless of whether you're in sales or development, services, finance, manufacturing, engineering, management or marketing, you possess skills crucial to your business and industry. And, although you probably don't like to brag about it, you're good at it. But you could be better, right? And by the way, who couldn't?

Frankly, we all could use improvement in our negotiating skills. Even if you're very good at negotiation, you may not know what makes you so good. That's why you got this book! Your peers might assume you have some kind of personal magic. But ask yourself: What is this magic and can I always depend on it? What happens when I wave my hand and no rabbit comes out of the hat? Is there a way that you could become systematic in your negotiations, more effective and efficient—particularly when you really need the rabbit to pop out of the hat? That's what we're here for.

This book will help you gain more predictability in your negotiations. Adding predictability to your negotiations will make your job easier. Trust us for now, but let us prove it!

YOUR GOALS

If we assume you are reading this book to be even better at negotiating than you already are, can we also assume that your job is already challenging? We don't want to make you work harder, because you are already working hard enough. Instead, we want to help you close your business faster with more confidence and fewer concessions.

But we don't want to put words in your mouth. Take a few moments to think about what *you* hope to learn from reading this book. For example, you might have some general goals, such as "to hone my negotiating skills". Or you might have more specific aims, such as "to get a better price and more value from Ms. Y at Company X". (By the way, Ms. Y has already read this book—so she's one step ahead of you!) We agree there is a time and place for almost everything in negotiation. We'll help you figure out the right time and place—where in the negotiation you may need to concede, for example. But you need to make concession decisions for the right reasons. We'll get to that, too.

We can teach you our method, but it's up to you to make sure you get what you want out of this book. That's why we urge you to write down your purpose for spending your **valuable** time with us. Notice the word valuable—the adjective form of "value"—is in boldface. You are going to see value mentioned a lot in the pages to come because it is a key concept in the negotiation process.

Consider the reading of this book as a deal you're making—a deal with yourself. As in all deals, you've got to begin by knowing what you want. What value are you looking for from this book? Jot down your goals in the Companion Workbook.

STOP if you are using the Companion Workbook.

Exercise 1-1: What value do I hope to get out of this book?

Here are some goals offered by our students in a recent negotiation seminar in Dallas, Texas. See how they compare to your own goals in learning the K&R Negotiation Method™:

"I want to simplify the complicated process of closing a business deal."

Sales Specialist

"I want to understand what salespeople are using as sales tactics. This will help me understand what they are *really* saying. Then, I can be a more effective coach. After all, we are all trying to drive the best deal."

Procurement/Purchasing Staff Member

"I want to learn how to negotiate from a point of disadvantage. We are not on a level playing field because our repeat customers are angry at us for past deals..."

Technical Salesperson

"I want to make more money and close deals more quickly."

Software Sales Rep

ROADBLOCKS TO SUCCESS

Consider these scenarios.

Scenario 1

> *You are the lead negotiator with the Big Deal Company. Your team is superb—all pros with a unified vision—and you are a skilled negotiator. You and your team are about to make a really great deal with the Fallen-on-Hard-Times Corporation. You know you have the leverage and the deal is a solid one for your team. (You'll learn all about leverage in Chapter 4.)*

> *Just as you are ready to get the Fallen-on-Hard-Times Corporation people to sign on the dotted line, the lead negotiator from procurement*

on their side leans across the table and says with a confident smile, "We can get this deal done if you give us an additional 10 percent discount." She's already asked this three times and there's nothing more you are willing to do for her. Besides, the business executive from the customer already agrees with your solution.

-or-

Just as you are ready to get the other company to sign on the dotted line, the lead negotiator from procurement leans across the table and says, "Unfortunately, your solution is not in our budget."

Scenario 2

You are up against Ed Duckbill, a well-respected buyer with more than a quarter century of experience at the negotiation table.

Ed is smart and well informed. He brings a wealth of experience to a negotiation. He's also great at using whatever tactics work. (You'll read all about tactics in Chapter 12.)

Unfortunately, some of Ed's tactics are really annoying. Several negotiators in the past have complained about Ed's aggressive behavior during a negotiation. When the talks get tough, Ed can become confrontational and hostile. He can even become downright vicious. The last time someone complained about Ed's malicious comments, he shot back, "If you can't stand the heat, get out of the kitchen. If you want me to buy your solution, just suck it up". You find his behavior tiring but you can deal with it because you know it's an act. However, several members of your team find his blustering, screaming and put-downs very distracting.

Scenario 3

Rex and his team at the Big Fancy Company did a spectacular job at the negotiation table. Their job, selling the company's software, required them to collaborate with their internal software developers. That's when the problem reared its ugly head. Rex's team cut a great deal externally only to see its internal divisions erode the company's ability to deliver. The negotiation team and the software developers sparred like children squabbling for their parents' attention. Tired of the battles, Rex finally asked the big boss to intervene. The response? "Work it out yourselves."

If negotiations were so easy, everybody would be successful. So, what's standing in your way? The roadblocks above fall into three main categories: *financial, tactical,* and *internal.*

Financial

- Our first scenario with the Big Deal Company and the Fallen-on-Hard-Times Corporation is an example of a financial roadblock. In this situation, the sticking point is money. In other words, "What else can you do for me? How much more can you shave from the price? How much more (free) support can you provide? How much of the shipping costs will you absorb? We only have X in our budget." You've heard all these gambits before. And you no doubt gnash your teeth when you come up against a financial barrier.

Tactical

- Our second scenario is a tactical hurdle. Ed Duckbill, our counterpart across the table, is using tactics to get a better deal. Tactics are not the sole territory of trained negotiators. Both professional negotiators and everyday customers use tactics. Because they are a common way that negotiators attempt to gain leverage and shift the advantage to manipulate bargaining, you will be hearing a lot about tactics in chapters to come.

Internal

- Our third scenario illustrates an internal conflict. Rex and his negotiation team have issues with the company's software developers. These conflicts prevent the negotiators and developers from being able to do their job effectively. In many instances, internal conflicts can be far more challenging to overcome than external financial and tactical hurdles.

Think about the hurdles you face as you negotiate. Then write the hurdles on the lines in your Companion Workbook.

STOP if you are using the Companion Workbook.

Exercise 1-2: My negotiation hurdles

TYPES OF HURDLES

For people in sales-related functions, hurdles to closing deals often include the following issues:

Financial

1. End-of-quarter pressure
2. Forced to sell price (not value)
3. Customer wants deeper discounts on renewal of your services or products
4. Customer threatens to cancel other business with you, if you don't lower price here
5. Qualifying the budget

Tactical

1. Unable to finalize, customer is always negotiating, no sense of urgency
2. Customer is confrontational or hostile
3. Customer wants price right away
4. Getting to the right decision maker or avoiding working with procurement

Chapter 1: About You and Your Value

Internal

1. Obtaining support of other people in your company working on the account
2. Management forces unnatural compromises just to close the deal
3. Lack of confidence
4. Team lacks synergy
5. Teammates contradict each other
6. Executives undermine negotiators' authority

We'll talk about ways to overcome these hurdles a little later in this chapter, and in detail in subsequent chapters. You will learn concrete techniques for overcoming these roadblocks.

YOUR PERSONAL TRAITS

No matter what value you hope to get out of this book, one thing is certain: you want to become a better negotiator. So, let's take a few minutes to find out what kind of negotiator you are *now*.

A really effective negotiator needs to have certain personal traits and business skills. What do you think those traits and skills are?

Let's look at personal traits first. Think of adjectives that describe the way people act. Here are some examples:

abrupt	emotional	egotistical
impatient	curt	rude

These might be the traits you possess as a human being, but they can be obstacles if you want to be a great deal maker. However, there are many traits you would like to have.

Chapter 1: About You and Your Value

In the companion workbook, list eight personal traits you think a really effective negotiator needs. These may not necessarily be traits that you possess. Rather, they are traits you believe would help a negotiator be more effective when working on a deal. Maybe you will even come up with more than eight traits. Write them all down in your workbook.

STOP if you are using the Companion Workbook.

Exercise 1-3: Eight personal traits effective negotiators need

Here are some adjectives that you may have listed. We have arranged them alphabetically, rather than in order of importance. Put a checkmark next to each trait that you already possess.

____ adaptable	____ direct	____ logical
____ analytical	____ experienced	____ neat
____ articulate	____ flexible	____ organized
____ attractive	____ forthright	____ patient
____ clean	____ honest	____ persistent
____ compassionate	____ humorous	____ persuasive
____ confident	____ insightful	____ team oriented
____ cool-headed	____ intelligent	____ tenacious
____ creative	____ intuitive	____ trustworthy
____ credible	____ knowledgeable	____ understanding
____ decisive	____ likable	
____ detail-oriented	____ listens well	

Chapter 1: About You and Your Value

Did you check them all? Of course, you didn't. (If you did, you might want to rethink your checkmark next to "credible".) We're not going to tell you that you should overhaul your entire personality. In fact, one of our axioms is "Be yourself. It's what you do best". But all of the above traits are important. For example, likability is important because, all things being equal, people would rather do business with people they like. Flexibility, adaptability and creativity make it possible for you to be effective even when something unforeseen occurs. Humor is invaluable because it helps both sides get through tense moments. Each one of the traits above is desirable in some way.

PATIENCE

So how do you fix your shortcomings? What can you do if you're not patient—one of our most important traits? Consider the following scenario:

> You are sitting at the negotiation table and it has been a l-o-n-g day. You have spent the last two hours stealing peeks at your watch...a lot of peeks! Are the people across the table ever going to close this deal? You want to go back to the hotel room and watch the basketball game or some ice skating. After all, it's championship time.
>
> One guy across the table doesn't even wear a watch. He smiles at you and says, "Gimme another five percent."
>
> "Oh, I just can't take this anymore," you think. "All right. You want five percent? You got it!" you say.
>
> The guy thinks for a few seconds. "My mistake, I meant ten percent," he says.
>
> Now what are you going to do? Your lack of patience has cost you, the price has eroded and the deal didn't close yet.

The K&R Deal Forensic

Several mistakes were made:
1. Patience to explain why the original price was just and rational was missing.
2. Instead of patience, we have an arbitrary or "unprincipled" concession.
3. The unprincipled concession leads to the other side requesting more and prolonging the process. After all, they probably don't feel confident they have reached your lowest price yet. Would you?

For many especially aggressive salespeople, patience is a difficult trait to acquire. But we *do* urge you to apply discipline if you are not naturally a patient person. Maybe you can take a break before responding and that will become habit. Without it, your emotions often get the best of you. And that erodes your deal, as in the following example from our client files.

> Our North American company was very close to closing a deal with a Japanese company. All we had to do was agree on a price. For our last scheduled meeting, our V.P. decided that he would join us. "I'll just sit in and help you wrap things up," is the way he put it. Unfortunately, our V.P. is not a patient person. We all sat down and, sure enough, the V.P. tried to "wrap things up".
>
> After a typically slow beginning and discussion, in a moment of silence, he blurted out, "We'll give you a 42% discount [which was 6% deeper than previous pricing]."
>
> The Japanese negotiation team said nothing. Why? This was a new offer. Based on their consensus decision-making style, this meant that the negotiators would have to present the 42% offer to their team and

management before they could give a response. There was no way they could agree to an offer on the spot.

Unlike the Japanese, most Americans and many Europeans and others are uncomfortable with silence. Once again, our V.P. took silence as a sign of their dissatisfaction with the 42% offer.

"OK," he said. "We'll offer you a 44% discount."

That's when one of us kicked him under the table. He kicked back.

More silence.

After a few more heartbeats, he said, "OK. For a deal this size I can go as high as 46%."

In a matter of seconds, the V.P. had eroded our deal by ten percentage points. That represented hundreds of thousands of dollars.

Yes, his lack of patience had gotten the better of him...and our whole company!

The K&R Deal Forensic

Several mistakes were made:
1. The V.P. acted without knowledge of the culture of the other side.
2. The lead negotiator may have thought the V.P. understood the culture and where the deal actually stood, but should have made sure before the meeting started.
3. The V.P., not intimately familiar with the deal, should not have intervened.
4. The V.P. had no patience and eroded the deal unnecessarily.

But what do you do if, even after you have tried and tried and tried, you still lack patience? Or what if you are just not likable? Or flexible? Or humorous? Does that mean that you can't negotiate? Is your career as a negotiator over? No. It's not. That's where teamwork comes in.

UNITED WE STAND, DIVIDED WE FALL

Teams are an essential element in successful negotiations.

Think of a small team...something everyone can relate to. You and your spouse are buying a house. While you are negotiating over price based on things that need to be repaired, your spouse exclaims, "I love this garage; it's the only one that fits my needs." What happened to your price negotiation?

A team can and should be your greatest asset. But an uncoordinated team can be your greatest liability. You need to get a team together that can stick together. And since you are trying to compensate for your own weaknesses, you don't pick the people who are exactly like you. Instead, you pick people who have the strengths you lack.

For example, if you're not willing to ask the difficult questions, get team members who are. Solicit advice from people who challenge your assumptions, who welcome confrontation. If you are not especially well organized or neat, get a team member who likes to straighten stacks of papers and put everything in alphabetical order. Give yourself the edge by assembling a team of players who complement each other's strengths and weaknesses. But then, you must manage that team. In Chapter 13, you'll learn how to assemble and work with a negotiating team, small or large.

STAY COOL

What if the other side has negotiators who are rude, inflexible and humorless? What if they are downright nasty? We talked about this hurdle earlier and promised you a solution. Remember? Consider this scene:

Chapter 1: About You and Your Value

The chief negotiator for Dewey, Cheatum and Howe is sarcastic and condescending. In the middle of a tense negotiation, he says:

"Gee. Harvey, I'd make this offer if I thought you and your team could understand it." He is a nasty piece of work. "Arrogant" and "obnoxious" come to mind. Plus, he fails at personal grooming. Add "offensive" to the mix.

What do you do?

You say to yourself and your team, "SO WHAT?" Yes, you would rather do business with people who are pleasant. Yes, you would rather do business with people who are not rude and arrogant. Who wouldn't? But if you're doing business with someone who is annoying or mean-spirited, stay cool and focus on the merits of the transaction. And, by the way, if you are dealing with someone who doesn't have good grooming habits or doesn't use deodorant, telephone negotiations may be more appropriate. We'll discuss those in Chapter 14.

KEEP YOUR FOCUS

Any worthy negotiator on the other side—although we like to refer to them as "business partners" or "customers"—will attempt to recognize and understand your strengths and weaknesses early in a negotiation. From that point on, they will try to take advantage of your weaknesses and get around your strengths.

Keep your emotions in check. As a negotiator, you should focus on the merits of the transaction. Ask yourself: "Does this deal make sense for both parties?" If the answer is "yes", then go about your business and make the best deal for yourself and your company. Remember, as long as the deal makes business sense for them also, they would look very bad to their peers and superiors if they didn't get it done. More on this in

Chapter 6.

YOUR BUSINESS SKILLS

What are your business skills? What are your sales skills? Are you an expert in intellectual property? Are you especially adept in personnel management? Are you a whiz kid with numbers? How are your legal and technical skills? Every business skill finds its way into the negotiation process. You may not need each skill in every negotiation, but you should be able to draw upon a deep well of business knowledge to be really effective.

Think of your training. Consider college classes and on-the-job training. Take into account the entirety of your business, too. Then list at least eight business skills or areas of knowledge that you think someone must have to be a really effective negotiator. Write these down in your workbook. If you can think of more than eight, list the extras, too.

STOP if you are using the Companion Workbook.

Exercise 1-4: Eight business skills really effective negotiators need

Here are some business skills that you may have listed. We have arranged them alphabetically, rather than in order of importance. Put a checkmark next to your areas of expertise.

Chapter 1: About You and Your Value

___ Accounting	___ Information Systems	___ Personnel
___ Contracts	___ Internet Utilization	___ Political Power
___ Customer Knowledge	___ Legal	___ Political Savvy
___ Customer Operations	___ Management	___ Procurement
___ Customer Service	___ Manufacturing	___ Project Management
___ Development	___ Marketing Methods	___ Sales
___ Distribution Methods	___ Mathematical Skills	___ Sales Contracts
___ Engineering	___ Negotiating	___ Sales Process
___ Finance	___ Experience	___ Technical Expertise
___ Industry Knowledge	___ Operations	

There's no way you checked them all! You can't know everything—no one can. But you probably already have a really good idea about how to handle any gaps in your knowledge. Consider the following:

> *John has absolutely no technical expertise. If somebody hands him a hammer, he says, "Hey, wait a minute! You're going too fast." But John has great negotiation savvy, and he anticipates beforehand that he needs someone with technical expertise on his team. Someone like Debbie, for instance. Now, when a technical issue comes up in a negotiation, John is a genius. He says, "Debbie will take that question."*

As we've already mentioned—teamwork, teamwork, teamwork! That's how you can overcome your individual weaknesses. There may be a lot of people on your company's payroll besides you. (And if there aren't, bounce ideas off a friend who can help.) If you are in the sales department, remember that your sales help pay the salaries of all the other people in the company! So, don't be shy about asking for help as you form your negotiations strategy. Because it is such a crucial element in negotiation success, we'll return to the importance of teamwork later in the book. In the meantime, remember these key points:

1. Make sure your team includes a variety of people. You can walk alike, you can talk alike, but you don't want everyone to think alike. If you do, your team won't be as effective.

2. Seek out people with the strengths you need. If you're not a tech person, be sure to get a technical expert on your team. If you're not a financial whiz, be sure to get a financial person for your team. Got it? Get it!

3. You must be able to have lively discussions internally, but be united externally. Your team must work as a *team* and present a unified face to the other side. Remember the spouse who loved the garage.

WHAT YOU LEARNED IN THIS CHAPTER

- Many traits are required for successful negotiations.

- Few people, if any, have all those traits.

- Patience is one trait we should all try to develop.

- Teamwork should supplement the important traits you lack.

- Get people with different personality traits on your team; use their strengths to compensate for your weaknesses.

- Concentrate on the merits of the deal, not on the behavior of the person on the other side.

- Identify areas in which you lack business expertise; then find experts in these areas to work with you on your negotiation team.

CHAPTER 2: THE ART AND SCIENCE OF NEGOTIATION

The famous detective Sherlock Holmes and his sidekick, Dr. Watson, are camping in the forest. They have settled in for the night and are lying beneath the night sky. Holmes says, "Watson, look up. What do you see?"

"I see thousands of stars," Watson replies.

"And what does that mean to you?" Holmes asks.

"I suppose it means that of all the planets in the universe, we are truly fortunate to be here on Earth. In a philosophical sense, it means that we should struggle every day to be worthy of our blessings. In a meteorological sense, it means we'll have a sunny day tomorrow. What does it mean to you, Holmes?" Watson asked.

"To me," Holmes replies, "it means someone has stolen our tent."

As this story illustrates, everyone has his or her own way of viewing reality. Some people, like Dr. Watson, view the world from a philosophical slant; others, like Sherlock Holmes, are more practical. To optimists, the glass is half full (as in, "Isn't this a beautiful night?"); to pessimists, the glass is half empty (as in, "The tent is gone."). People are similarly divided over the issue of negotiation: Some people see it as an art, others, as a science. Can it be both?

WHAT IS NEGOTIATION?

 Before you learn how to hone your negotiation skills, let's make sure we're all on the same page. Take a moment to define **negotiation.** Write your definition in the workbook.

STOP if you are using the Companion Workbook.

Exercise 2-1: I define negotiation as...

We define negotiation as "interaction or discussion between people to reach agreement". A successful negotiation means reaching agreement on terms that satisfy both sides' goals to varying degrees.

Some people view negotiation as a shifty process. Nothing could be further from the truth. The world of commerce today is dependent on people making business deals that create value. This adds to the world's economic well-being. Thus, negotiation is a very positive process. If you do a good job negotiating a deal, you create a relationship that lasts. The deal delivers something of value to both sides that they otherwise would not have had.

This does not mean that negotiation is always a peaceful, cooperative process. By nature, the process has adversarial aspects. After all, most of us have different views of the value we should get. And most of us are somewhat competitive. If negotiation was an exact science and value was exactly calculable, we would all agree on terms immediately and there wouldn't be any need to negotiate. We could all go home. However, negotiation is ultimately a positive process because its goal is positive: to achieve a result that meets as many needs as possible and improves the respective positions of the participants.

TYPES OF NEGOTIATIONS

Negotiations occur in different situations. These include:

- Personal negotiations

- Internal business negotiations

- External business negotiations

A *personal negotiation,* for example, might involve discussing with your spouse or significant other what you are going to do Saturday night. Will it be the movies? Perhaps dinner with friends? Maybe bowling? You have a personal negotiation when your young child demands a toy/ice cream/merry-go-round ride during a shopping trip to the mall. And we don't even want to discuss personal negotiations with your teenager!

Internal business negotiations, in contrast, might involve fighting for budget dollars, or asking for a raise, a promotion, or a choice perk, such as the privilege of flying business class. Internal business negotiations don't have to concern just you. As we mentioned in Chapter 1, some of the most difficult negotiations are internal ones in which you are deciding which policies or steps to take externally to be successful. Many people in business never get involved with external negotiations. However, they are instrumental in shaping policy through internal negotiations. That policy then gets externalized through negotiations with other companies.

External business negotiations involve dealings with people outside the company. You might be selling or buying a product or a service, for example. You might be negotiating a trade agreement or a shipping deal, or arranging lodging and food for a major company retreat.

Of course, there are other types of negotiations whose specifics are beyond the scope of this book. These include litigation settlement discussions and political negotiations, such as peace talks.

THINKING ABOUT VALUE

We introduced the concept of value at the start of Chapter 1 because it is a central element in the K&R Negotiation Method™. It's important for you to understand your own needs—and the needs of the people on the other side of the negotiating table—before you can negotiate a deal that works best. Yes, in most cases, your needs and their needs will overlap. But never forget that you and the folks you're dealing with are *people*, not abstractions. People have emotions. The following scenario illustrates our point.

> *You're negotiating to buy a large quantity of electronic devices for your company. You know that this seller usually offers a 20% discount, supposedly non-negotiable. Nonetheless, your boss insists that you try to negotiate a 25% discount.*
>
> *"Listen," you slyly tell the seller, "I'd love to close this deal right now. If you could only give me a 30% discount, I'd..."*
>
> *"Done!" says the seller.*

You got more than your boss wanted, yet you know that you could have done even better. How do you know? It was too easy. The concession from the seller appeared as if he had more to give—but it appeared of little value. Surely this deal is in your company's best interest. But on a personal level, you feel cheated! What would have happened if you'd asked for a 40% or even a 50% discount? Mentally, you're kicking yourself. And that can hurt!

Although you got a great deal—even better than you had anticipated— you feel victimized. Why? Because your emotional needs were not met. You, as a customer, would have felt better if you got a **principled**

concession. And the seller would have been better off giving you a good reason for the concession so you wouldn't feel like a victim. That is, you need to know *why* the other side would budge from their usual 20% discount. You want to hear a good reason. By "good", we mean a reason based on a credible business rationale related to **value.** Maybe you would have felt better hearing something like, "Well, since we would really like to have your business now *and* in the future, we will give you a 25 percent discount, *if* you agree to buy so-many devices over X number of years."

This scenario suggests that most people feel better about their decision to complete a deal when the concessions in the process are rational rather than arbitrary, even when the arbitrary concession may work financially more in their favor. The concession received was a 30% discount, yet you felt cheated. The principled concession was 25%. Thus, the unprincipled concession was a 5% lower price than the principled one. But even getting 5% less, you would have felt like a negotiating whiz instead of someone who just got ripped off. This is a key concept in negotiations, so we'll come back to it many times, especially in Chapter 6.

QUANTIFYING VALUE

In Chapter 1, we asked you to think about the value you want to get from this book. Look back over your answers. Ask yourself: "Are my answers specific or vague? Can I **quantify** them?" Remember that we told you to consider your reading of this book as a deal. You contribute your valuable time, and we must offer something in return. Obviously, we don't expect you to come up with a dollar amount. We don't expect you to say, "Reading this book will be worth exactly a million and a half dollars to me in the next year." (It might be worth more, by the way.) But we want you to be able to think of specific situations when you wish you had struck a better deal, and, more importantly current or upcoming situations where you can apply what we discuss to get what you are pursuing. That is what we are after.

Remember: Whether you are a buyer or a seller, you should have a quantifiable value argument in mind.

Let's look at another familiar scenario. Bet you've been involved in this one, too.

> *It's April 4. You're trying to sell your product or service to a customer you have dealt with before. Unfortunately, you have "trained" this customer to wait until the end of the quarter to sign on the dotted line, because that's when you are usually ready to offer an additional 10% discount just to close the deal. So, of course, the customer is thinking: "I'm not going to do anything until her end of quarter, June 30." She listens to your sales pitch politely, but tells you, "Let me think about it. I have to toss the idea around with some key people here. Give me some time, and I'll get back to you."*

Well, you really didn't give her any ideas to toss, did you? What you failed to do was give this customer a **compelling argument** to buy now, today, immediately. You did not create a sense of urgency. You'll be hearing a lot about creating a compelling argument in this book. It's covered in depth in Chapter 9.

What's the best way to create a compelling argument, a reason to buy now? **Quantify the value.** Explain to the customer how her company is losing X amount of money every single day it *doesn't* use your solution or service. Say that you hate to see their company losing that kind of cash. By the way, "that kind of cash" obviously needs to be more than they could save by waiting for the discount. Otherwise their best decision would be to wait for a better price.

ART VERSUS SCIENCE IN NEGOTIATION

Art

The definition of art is "using imagination to create beautiful things". How does this apply to the art of negotiation?

Science

The definition of science is "a systematic activity requiring study and method, knowledge acquired through experience." What science is involved in negotiation? Using the workbook, write down your definitions.

STOP if you are using the Companion Workbook.

Exercise 2-2: Art and science applied to negotiation

You may have heard the phrase "the art and science of negotiation". It's a cliché all right, but clichés have staying power because they contain more than a grain of truth. Think about the two parts of this saying: art + science. We came up with these characteristics. See how they compare with yours.

Negotiation is an art because...

- each deal is special in its own way.
- motivations vary and cannot always be predicted accurately.
- goals are usually different.
- emotions are involved.
- people are unique - personalities are different.
- the formula is not set.

- creative solutions are often required.

Negotiation is a science because...

- it involves method and process.

- costs, prices, and margins can be determined.

- there are identifiable alternatives.

- the competition can be studied.

- there is a bottom line.

- one's motivations and goals can be determined.

- the opponent's motivations and goals are discoverable.

- people's behavior is often predictable.

- predictability of behavior is based on psychology (a science).

- outcomes are often predictable.

NEGOTIATION AS AN ART?

Consider the following story.

> *We were about to enter into settlement negotiations with a company, IBID, that had sued our client, Tinco, for the third time. The owners of IBID were considered outstanding citizens in their community. They wanted vindication, not money. Unfortunately for our side, the lawsuit also had merit. For this first negotiation, Tinco's president, Clyde, was present. A rancher, Clyde was always dressed in cowboy boots, hat, and other Western gear.*

> *From previous discussions with IBID's owners, we knew they were very intense people. As we were going up the elevator in the Manhattan office building, Clyde looked out of place. I asked him to pause at the threshold when we entered the conference room and look at his boots.*

Chapter 2: The Art and Science of Negotiation

When we got to the conference room, Clyde paused and did his best Clint Eastwood imitation looking down at his boots and up again. IBID's owners couldn't help but laugh at seeing this cowboy in mid-Manhattan. The tension was relieved, they took an immediate liking to Clyde, and we settled in two weeks.

Negotiation is an art because it involves creativity to solve problems and bring closure to deals. You often have to think "outside the box" to create a value argument. That's because negotiations have so many variables. There are variables relating to products, services, and the alternatives available to the parties. Then there are the human variables. The person you're negotiating with can be loud or quiet, assertive or passive, credible or unprincipled, experienced or untried, emotional or placid, quirky or predictable. Or, he or she can have all these qualities. As a matter of fact, so can you! On top of that, company histories, philosophies, and practices vary, as does their size and market position.

All of these different conditions, behaviors, and personality traits make it a challenge to work with people. Perhaps you use humor to defuse a tense situation. Maybe you're one of those calm people who let the pressure roll right off your back.

NEGOTIATION AS SCIENCE?

Consider this riddle:

Question: Why are there so many Smiths on Facebook?

Answer: They all have accounts.

Negotiation can be considered a science because you can deal with empirical data. For example, *there are so many Smiths on Facebook because they all have accounts.* Can't argue with that.

There are many empirical situations you likely encounter every day. For example, if you're creating a product, you design the specifications. You know these specs create specific functionality. You build a prototype; you test it. You observe that the product works as you assumed it would (or that it didn't, but then it's back to the drawing board). You also know how much the parts cost, what development and manufacturing resources you need, how much you expect to sell, and what your profit margins should be. In short, you're armed with a lot of hard-and-fast data. This data enables you to price the product appropriately within reasonable parameters. Nice and almost scientific. (Note: Value forecasts are based on assumptions. If assumptions are wrong, so is everything else and the science is flawed.)

Remember that science is defined as "a systematic activity requiring study and method; knowledge acquired through experience". In negotiation, there is a lot of predictability in peoples' behaviors. We understand that from experience. If we set our emotions aside and assess "how this person is likely to react if I do that", more often than not we are going to be right. That kind of calculated approach alone can increase our success rate in negotiations. It is this predictability that is critical to effective planning in the negotiation process.

NEGOTIATION AS AN ART <u>AND</u> SCIENCE!

We see the process of negotiation as both an art and a science (you saw that conclusion coming, didn't you?). Negotiation is not an exact science like mathematics. The K&R Negotiation Method™ is based on scientific principles that we have refined, which increases predictability, but it factors in the human element and lets us deal with surprises and fashion solutions creatively.

The concepts we discuss have many psychological and environmental elements that could lead to exceptions. When we negotiate, we are dealing with the realm of probabilities, not certainties. As a result, when planning for a negotiation, you should always keep the K&R Negotiation Method in mind. But do not be afraid to exercise exceptions to our general principles—as long as you do so *consciously.* That means that you:

- Understand our general principles

- Take the risks inherent in not following the general principles of the K&R Negotiation Method™

- Understand the "down side" of taking those risks

- Recognize the trade-off you are making to gain a greater benefit

And while we are on the subject, it's time to throw out the misconceptions and myths that may be holding you back. Perhaps you're less comfortable operating outside empirical situations. You might feel uneasy being creative in interpersonal situations. Maybe your lack of experience translates to a lack of confidence.

Here's a new paradigm: You can be a more effective negotiator if you consider negotiation as both a science and an art. Here's how to do it.

Use science to

- Take the time to gain knowledge

- Support your credibility

- Improve confidence

- Make you more believable

- Form the foundation to articulate value to the other side

- Help you communicate persuasively

- Intelligently predict the likely outcome based on your experience and that of your peers

Use art to

- Look for creative alternative solutions

- Be open-minded to problem solving

- Fulfill different motivations

- Apply intuition in judging motivations and personalities

- Deal with different personalities

- Stand back, set your ego aside, and allow the experts on your team to take charge

The K&R toolset will help you do these things. The following scenario illustrates the combination of art and science in the K&R Negotiation Method™.

> *We were negotiating for a client who was doing a technology deal with a Mideast company. Our client would be supplying both hardware and software that the customer would resell as part of their own customer offerings. They would rely on our client for dedicated technical support for the technology. In the previous three years, the customer had very volatile revenues and earnings.*

> *When we represent a supplier, we like to include a contract term in which the buyer agrees to revenue or volume commitments, especially when the buyer requires dedicated supplier resources for support. That usually means the agreed-to prices are based on achieving milestone revenue or volume levels. This term serves at least three purposes:*

> *First, it gives some protection to the supplier, who is committing resources to the relationship. It allows the supplier to reduce its commitment or terminate the contract if the buyer doesn't purchase the agreed-upon amount. There are those rare times that terminating a relationship becomes a better business decision because of the costs*

associated with keeping the relationship alive. Second, lower prices serve as an incentive to drive revenue and volume. If the volumes don't happen, neither should the reward of lower prices. Third, the supplier can better manage its resources by obtaining financing based on revenue guarantees from its customers. This enables a supplier to have more efficient component or subcontract service acquisitions and be a more reliable supplier.

Early in the negotiating process it became clear that the buyer would not agree to any type of volume or revenue commitment. Two very rational arguments were made to support their position. They said, "Due to our past volatility, there is still too much uncertainty to make commitments, even though we believe we are on the right track." They added, "But even if times get tough again, we would still need you as a supplier or we might go out of business. After all, we can't generate revenue if we don't have a product to sell."

This was a big deal: $15 million per year. Our client, the would-be supplier, wanted to do this deal. How would we approach the problem? Harvey, leading our team, got creative and tried to obtain a sole source agreement in lieu of a volume commitment. The customer's negotiating team wouldn't go for it because they did not always control their division's final buying decisions. They also had pre-existing competitive installations that were incompatible with the supplier's product line. Again, they provided a rational—scientific!—argument for not going sole source.

We made our arguments again, to demonstrate the importance of this issue to us. We said, "As a supplier, we are obligated to perform. For us, it is costly to support a customer." Harvey settled back in his chair and asked, "How do you view our relationship? You don't want minimums and you don't want a sole source relationship. We need some sort of comfort that you are going to buy our solution."

The goal? To see if the buyer could come up with a creative solution to the dilemma. If the right idea came from them, we all knew that it would be more acceptable to them than the same idea coming from us.

Chapter 2: The Art and Science of Negotiation

Their response? "Absolutely! We regard you as our primary supplier. If one of our divisions comes in with a requisition that's not for your products, we'll make every effort to persuade them that your company is the way to go. Furthermore, if the decision is strictly economically-driven, we'll give you the right of first refusal to match the price."

We replied, "That is an excellent commitment to the relationship. Based on what you said, we can include some terms in the contract that will work for both of us." Our team wrote up what the buyer had said, word for word. Since the customer saw their own words in print, how could they disagree? They did not. We then defined "primary supplier" as 75% of new orders placed by volume. We moved on to close the deal and cement the relationship based on a primary-supplier commitment.

The K&R Deal Forensic

Here is what worked:
1. Patience and listening helped us gain knowledge about both sides.
2. Using that knowledge, we solved both sides' problems with a creative solution. The art and science of negotiation at work.

How do you solve problems when the hypothesis—the empirical fact that your initial terms are part of your initial price and belong in your contract—isn't working to close the deal? You look at what problem you are really trying to solve (more on that later). Then you come up with a creative solution—you marry the art and science of negotiation.

OVERVIEW OF THE K&R NEGOTIATION METHOD™

We have combined the art and science of negotiation to create a winning formula for success. Our process is based on K&R's Six Principles™. You will be reading a lot about these principles in the chapters to come. They're covered in detail in Chapter 6. Here's a sneak preview:

K&R's Six Principles™

1. Get M.O.R.E.—Preparation is key to a winning negotiation.

2. Protect your weaknesses; utilize theirs.

3. A team divided is a costly team.

4. Concessions easily given appear of little value.

5. Negotiation is a continuous process.

6. Terms cost money; someone pays the bill.

Look back over these Six Principles. Based on what you have read already, what words within the Six Principles do you think are most important? Circle them in your workbook now. Then write a brief explanation of your choices.

K&R'S SIX PRINCIPLES™

STOP if you are using the Companion Workbook.

Exercise 2-3: K&R's Six Principles

Here are the words we see as most important. (The key words are in bold type.)

1. Get M.O.R.E.—**Preparation** is key to a winning negotiation.

So, you want to buy a new car, a new dishwasher, or $10 million worth of software for your company. No matter how big or how small the deal, do the legwork first. In the case of buying a car, find out the dealer cost and the cost of each option. How much does the dealer *really* pay for that upgraded interior package? Ditto on the dishwasher and even more so on a huge corporate deal.

But no matter how big or how small the deal, **preparation** is key to a winning negotiation. (We'll discuss what M.O.R.E. stands for in Chapter 6.)

2. **Protect** your weaknesses; utilize theirs.

You can't add five and five without looking at your fingers? Or, maybe you are quite good with numbers but you don't know much about contracts? What do you do? The side that addresses its weaknesses has an advantage that translates into leverage (we will discuss in Chapter 4 and throughout). Leverage gives us the ability to obtain better terms or prices. Conversely, weaknesses give an advantage to the side that doesn't have them. Assemble a tightly knit team that collectively has the skills you need to **protect** your weaknesses and utilize theirs. You'll read more about this later.

3. A team **divided** is a **costly** team.

How many times have you been involved in this scenario?

> *Junior corners his mother and asks for $50 to take his new girlfriend out on a date. "$50!" his mother says. "Are you nuts? You never take out the garbage, your room is close to being condemned, and we certainly do not want to talk about your grades. If you want to take a girl out on a date and spend $50, you better get a job."*

> *Junior is a big-time slacker when it comes to chores and studies, but he's whiz-bang when it comes to negotiating. Before his father gets home and his mother can fill him in on the discussion, Junior gives Daddy a call on the cell and says, "Father dearest, remember when you were my age? Remember how great it was when you finally got a date with a really great girl? I got a date with the most popular girl in school, but I don't have the money to show her a good time. If only I had $50, I*

could take her to a club and buy her a really nice dinner. Mom just doesn't understand how we men feel. Whaddaya say, Pops? I can always count on you."

Now, "Pops" happens to be quite a sharpie, but he is in the middle of a tough business discussion and he's distracted. As a result, he does not suspect that Junior is playing "divide and conquer". Guess who wins this one?

Children are experts at playing one side against the other. A team **divided** is a **costly** team. Your customers can be pretty good at it, too. Work as a team and you stand a much better chance of success.

4. **Concessions** easily given appear of little value.

Consider this story. It's from our files.

The buyer asks the salesperson to give an additional 15% discount (for a total of 40%) and offers to sign the deal now. The salesperson quickly says, "OK." The buyer feels as though she could have gotten more and balks at signing, saying, "I'll have to check with my boss." Alternatively, the salesperson may say: "Please tell me why you feel our solution is worth 15% less." The buyer may say, "Well, we don't believe that two of the five functional elements could be implemented within the next six months." If this statement is true, it is a credible argument. The salesperson now has the opportunity to make a principled concession, a concession related to value. After taking a break to do some analysis, the salesperson may respond, "We understand your point, and the fact that certain functions cannot be implemented right away is worth 7.3 percent. Let me explain why." This gives the salesperson the ability to satisfy the buyer while maintaining a credible, principled position. This approach builds relationships and the difference between the 15% and the principled concession is revenue and profit that you keep!

However, the best principled concessions are not ones you give, but ones that you induce from the other side, based on the value of what you do for them. Usually both are involved in the process.

5. Negotiation is a **continuous** process.

Check out this true tale.

> *A Telco customer engaged an IT Service provider in a $4.1M fixed-price custom application development project. The application was defined by a client requirements document prior to signing, which is not specific in a number of places.*
>
> *In the first two months of the project, Telco requests small changes to the specification, which IT Service's delivery team implements. In month 3, Telco makes an additional request that IT Service estimates would require significant additional resources for the project. As a result, IT Service requests that the specification be "formally" amended for an additional payment of $500,000. Telco refuses, saying that its request is just a clarification like its previous change requests and should be part of the project scope. Now we need to return to the negotiation table.*

Not all cases are this extreme, but you can expect that people will have different memories and interpretations of facts and **negotiation** will be a **continuous** process. In the technology industry in particular, many deals are signed before all the facts can be known and while technological progress is taking place. Very often, licenses involve continuous maintenance and upgrades. It's not a simple buy-sell situation. By their nature, deals are continually negotiated. That is why we have individual attachments or statements of work for almost all contracts. It's also why we have amendments, change of control, and other governance provisions that anticipate continuous negotiations.

6. Terms cost money; someone pays the bill.

According to the old saying, there's no such thing as a free ride. The price may not come out of your sales commission, but it surely will come out of someone's budget. Terms affect the financials of a transaction, whether it's a price or a budget issue, or usually both. For example, the price will be different for a one-year warranty versus a five-year warranty (a four-year warranty extension for an electronic item at major retail stores often costs 25% or more of the product price). This means that not only will the price be affected but so will the resources and budget allocated to support the additional warranty term.

In the chapters to come, you will learn about each of these six principles in more detail.

Chapter 2: The Art and Science of Negotiation

WHAT YOU LEARNED IN THIS CHAPTER

- **Negotiation** is "interaction or discussion between people to reach agreement".

- A **successful negotiation** means reaching agreement on terms that satisfy the goals of all the parties to varying degrees; each company improves its marketplace relationship or position.

- Concessions should always be principled. More on this in future chapters.

- Always try to quantify value. Don't assume the other side will understand the value. Don't be vague about it. Spell it out.

- Try to give the customer a **compelling argument** to buy now, today, immediately.

- Negotiation involves both art and science, both facts and creativity.

- K&R's Six Principles™:

  1. Get M.O.R.E.—Preparation is key to a winning negotiation.

  2. Protect your weaknesses; utilize theirs.

  3. A team divided is a costly team.

  4. Concessions easily given appear of little value.

  5. Negotiation is a continuous process.

  6. Terms cost money; someone pays the bill.

CHAPTER 3: EFFECTIVE COMMUNICATION

Question: What do you call a boomerang that doesn't work?

Answer: A stick.

You learned in Chapters 1 and 2 that to negotiate an effective deal, you must understand your needs as well as the needs of the people on the other side. To get your message across, you communicate. Sounds easy, but we all know it's not. In fact, communication is a far more complex process than you might have imagined.

WHAT IS COMMUNICATION?

Communication is the process of creating meaning. Sometimes people think of communication as a series of isolated actions. Person A says one thing; Person B says another. Sorry, wrong number—that's not how communication works.

Communication involves four parts: sender, message, receiver and response. The way we interpret speech and communicate in specific situations depends on a wide variety of variables. These include our personal experiences, our mood at that moment, and our overall values. Communication is an ongoing process with variable conditions and outcomes.

- Communication arises from **context.**

Context is the time, place, and circumstances of the communication experience. Meaning is constructed as a social process, embedded in the context of the interchange. Context influences what we say, how we say it, and the way others understand it. For example, the meaning you take from a colleague's praise for your new outfit depends in part on when the praise is offered. If it comes at the beginning of the day, it has one meaning. "What a nice way to start the day," you might think. But if your colleague asks to borrow your car right after she compliments your outfit, you're likely to construe a different meaning. "She only said something nice to get a favor," you think. Communication changed because of the time and circumstance.

- Communication is **symbolic.**

Words and actions carry **symbolic** overtones or meanings greater than what they appear to be. First, think about words. For example, to a person who speaks English, "pain" means aches, discomfort, and suffering. To a French speaker, however, the word carries no such meaning. In French, "pain" means bread. Every language has a grammar or system of rules that tells us how to use these symbols to communicate with each other. Understanding and following these rules enable us to communicate with each other.

Words are not the only thing to consider when we communicate; actions also have symbolic meanings. For example, making a circle by placing your thumb and forefinger together means "A-okay" in English. It's generally accepted as a symbol for agreement in the North American culture, but in many Latin American and other cultures, it is an obscene gesture.

Consequently, actions as well as words can create embarrassing and even dangerous communication problems.

Ever hear this classic anecdote?

Chapter 3: Effective Communication

Two women approached Calvin Coolidge. The president spoke so rarely that he had earned the nickname "Silent Cal." One woman said to the close-mouthed president, "Mr. Coolidge, I just bet my friend that I could get you to say three words."

Said Coolidge, "You lose."

Communication is a social process of making meaning. Communication is not separate, isolated acts. Communication would be easy if people asked and answered questions in predictable ways and if you could ignore the context. It would be even easier if every exchange was scripted in advance and we just read from the pages. But that is not the case.

IT'S WHAT YOU SAY *AND* HOW YOU SAY IT!

Often—in negotiations as in our personal lives—it is not what you say, but how you say it. Our interpretation of words in a specific situation and the way we respond to them depend on our experiences, values, and emotions. Sometimes people use different words to mean the same thing. However, many words carry loaded overtones, **connotations,** different from their dictionary meanings or **denotations.**

For example, the words "cheap" and "thrifty" have the same denotation (dictionary meaning): frugal. But wouldn't you rather be called thrifty than cheap during a negotiation? (Or at any time, for that matter!) That's because the two words have very different connotations or emotional overtones. Cheap has a negative connotation of "tight-fisted", while thrifty has the positive connotation of "wisely economical". In a similar way, the words "stubborn" and "resolute" both mean persistent, but stubborn carries a negative connotation of "obstinate" and "pigheaded", while resolute carries a positive connotation of "purposeful" or "resolved".

As a negotiator, it's important to be sensitive to the use of words, otherwise you run the risk of offending someone. More importantly, their behavior toward you may change to the negative and you won't have a clue why!

You are not likely to use all the following words in a negotiation. These words have similar denotations, but not the same connotations (emotional overtones). Sort the words using the chart in your workbook.

trashy	*economical*
cheap	*shoddy*
moderate	*competitive*
common	*budget*
flimsy	*reasonable*

STOP if you are using the Companion Workbook.

Exercise 3-1: A house is not a home

Here's how we arranged the words in the exercise:

Positive Connotations	Negative Connotations
Moderate	Cheap
Reasonable	Common
Budget	Shoddy
Economical	Flimsy
Competitive	Trashy

Remember this exercise. It will help you choose words carefully as you negotiate.

COMMUNICATING RESPONSIBLY

It's true that speaking itself may not be difficult, but we have learned in this chapter that communication is not the same as speaking. In a negotiation context, we are always trying to bring people closer to our way of thinking. We cannot do that unless we understand what they mean and have our listeners understand what we mean. Training and practice can make an enormous difference in our ability to make our meaning understood and our ability to understand what others are saying.

Communication is also influenced by what we want to perceive. In many cases, we assume that other people feel the same way as we do about a given issue. As you've probably discovered the hard way, this is not always the case. We also tend to tune out messages that are unpleasant, threatening, or disturbing. Instead, we take in messages that reinforce our sense of well-being or tell us what we want to hear. Read the statements in the workbook and check which ones you think make someone a responsible communicator.

STOP if you are using the Companion Workbook.

Exercise 3-2: Are you a responsible communicator?

Every statement in the workbook exercise is correct.

The process of conveying meaning is very complex. It is often easier to misunderstand people than it is to effectively communicate with them. Even the most skilled communicators can benefit from training and practice.

WHAT *NOT* TO DO: ERRORS IN LOGIC

Faulty logic can demolish the most carefully constructed value argument. Bad logic is one of the surest ways to lose an argument. Below are the most common "logical fallacies"—everyday errors in reasoning.

OVERSIMPLIFYING THE ISSUE

When speakers oversimplify the issue, they twist the truth by presenting too narrow a range of possibilities. For example:

> *Here, we have a clear-cut choice between a plan that will result in a huge boost in international sales or a plan that will result in an economic disaster both at home and abroad.*

Are the two sides of the issue really that clear-cut? Are there no possible negative results of the first plan or possible positive outcomes to the second? The flaw here is the use of fabricated statements. It seems unlikely that the two options are that obvious. Unless the negotiator can back up the assertion with convincing details, everyone in the room is likely to shake their heads in disbelief. The argument is not valid. And it gets worse for the negotiator. Credibility is eroded. The negotiation just got more difficult. And the deal will probably take longer, if it's done at all!

BEGGING THE QUESTION

In this type of logical error, a position that still needs to be proved is stated as though it has already been proved. For instance:

> *The question we must resolve is whether you should buy the product now or at the end of the quarter.*

This position and question, often referred to as a "trial close", are credible only when you have already made a compelling value argument that the customer should buy the product. Otherwise, the real issue is *not* whether the customer should buy the product now or at the end of the quarter. The real issue is *whether* the customer should buy the product at all. The speaker has avoided having to prove the real allegation by assuming it as a fact. Taking this approach without credible value arguments will seriously affect your credibility in negotiations.

USING MISLEADING STATISTICS

Leonard H. Courtney once said, "There are lies, damned lies, and statistics." Misleading statistics are usually true but do not prove what the speaker claims.

For example, you may have heard the advertisement that four out of five dentists surveyed endorse a certain brand of toothpaste. Over what: not brushing? All you know for sure is that five dentists were surveyed—not fifty, not five hundred, not five thousand; you don't know the actual number of dentists in the survey. And because these dentists may not be typical of the entire population of dentists, their endorsement may not provide an accurate representation of dentists in general.

USING FALSE CONCLUSIONS

There is a Latin phrase which means "after this, therefore because of this". This type of logical error is the mistake of concluding *because* from *after*. Here's what we mean.

> *During the new sales director's first year, the sales revenue declined 10%. Should we keep someone who cannot grow or even maintain sales?*

The fact that the sales revenue declined *after* the sales director took the position does not mean that it happened *because* she has the position.

Perhaps a key product failed testing. Perhaps there were no new products issued that year. Or competitors might have brought out a superior new product. The speaker should show that the events are indeed linked by a cause-and-effect relationship.

BACKWARD REASONING

This logical fallacy assumes that people belong to a group because they have characteristics in common with that group. Consider this example:

> *Sales directors are always proposing larger commissions. Seymour Miles is proposing a larger commission. From this we can conclude that Seymour Miles is a sales director.*

Clearly, people other than sales directors have proposed larger commissions. For example, what salesperson does not want a larger commission?

USING FALSE ANALOGIES

False analogies are misleading comparisons. The correspondence does not hold up because the items or people being compared are not sufficiently alike. For instance:

> *A good negotiation is like well-baked bread. Both have quality ingredients.*

No such luck! Negotiations and bread may have a few surface similarities, but the details in a negotiation and relationships between parties are far more complex than what goes into a loaf of bread. The rules for success in a negotiation are often infinitely more intricate than in baking bread, though some bakers may disagree.

P&L: DO YOU HEAR WHAT I HEAR?

Successful communication is like a boomerang—it comes right back at you. This is true because effective communication is a process of "P&L". Bet you're thinking "Profit and Loss". Of course, effective communication *does* result in more profit than loss. (Otherwise, who would bother?) But effective communication rests on the cornerstones of another type of P&L: patience and listening.

Here's what we mean:

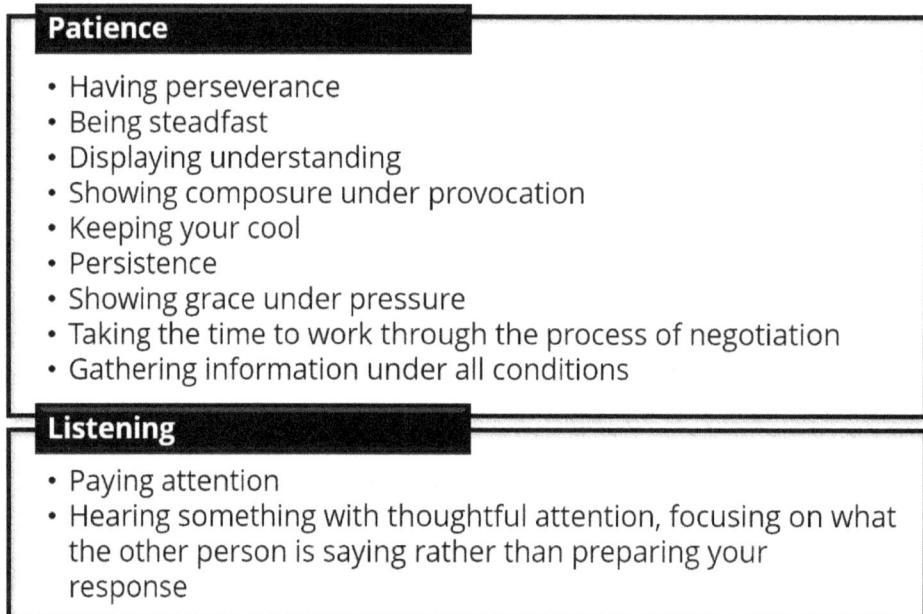

Patience

- Having perseverance
- Being steadfast
- Displaying understanding
- Showing composure under provocation
- Keeping your cool
- Persistence
- Showing grace under pressure
- Taking the time to work through the process of negotiation
- Gathering information under all conditions

Listening

- Paying attention
- Hearing something with thoughtful attention, focusing on what the other person is saying rather than preparing your response

Figure 1: Patience & Listening

Ultimately, to have good P&L (profit and loss), you must practice P&L (patience and listening). Why? Because it's patience and listening that enables you to gather information that you will use in articulating your persuasive arguments in negotiations.

Chapter 3: Effective Communication

LISTENING SKILLS

Keep in mind that speaking is a two-way process. It involves not only making contact with the audience, but receiving feedback from them, as well. It's not enough to be a good speaker; you also have to be a good listener. There are three main kinds of listening:

- Empathic Listening
- Informational Listening
- Evaluative Listening

Let's look at each one in detail.

1. *Empathic Listening*: This type of listening provides emotional support to help the speaker come to a decision, solve a problem, or resolve a situation. Emotions are the focus more than reason or ethics. As an empathetic listener, you can restate the issues, ask questions, and critically analyze the issues. Your intention here is not to decide for the speaker. Rather, it is to support the speaker's own independent decision-making process.

You can do this by inviting the speaker to express all ideas and feelings freely. For example, if a customer is telling me about company problems, I might say, "I understand your problems. We have heard of folks who have similar issues. Please tell us how you have dealt with your issues. I am sure a company as good as yours has had some good solutions." This approach allows a customer to speak freely as you provide emotional support. This makes a customer feel more comfortable and helps you gather more information to use in your value arguments later.

2. *Informational Listening*: In this kind of listening, the listener gathers as many facts as possible. The focus is on accuracy of perception. This is the type of listening you do when members of the other side in a negotiation ask you questions. It's the type of listening you need to do before responding, to make sure that you understand the question, task and issues. For example, you may say: "Please help me understand what you are looking for here..."

Informational listening demands that you focus on specific details, distinguish between different pieces of information, and organize information into a meaningful whole. You hear something a customer says and you want to gather additional information. You can do this by asking follow-up questions to the information you just received.

3. *Evaluative Listening*: With this kind of listening, you weigh what has been said to see if you agree with it or not. Start the process with informational listening to make sure you have all the facts. Once you understand the issues, you can then evaluate them and make decisions based on the facts, evidence, and speaker's credibility.

Evaluative listening is most helpful for negotiators in decision-making situations and confrontational positions.

HOW'S THAT AGAIN? POOR LISTENING HABITS

Being able to listen well is an invaluable skill for effective negotiators. Nearly everyone has some bad listening habits that can be overcome with awareness, practice, and training. Here are three of the most common bad listening habits:

1. *Pseudo listening* occurs when you only go through the motions of listening. You look like you're listening, but your mind is miles away. Correct this error in listening by really focusing on what the speaker is saying. Tune out distracting thoughts such as what you want to eat for lunch, what you're going to watch on television that night, and that

you would rather be outside playing golf!

2. *Self-centered listening:* Ever mentally rehearse your answer while the other person is still speaking? That's self-centered listening. It's focusing on your own response rather than on the speaker's words. Correct this listening fault by letting the other person speak before you begin to frame your answer. (We know this is really, really difficult, but it is crucial to a successful negotiation.)

3. *Selective listening* happens when we listen only to those parts of a message that directly concern us. For instance, during a negotiation you may let your mind drift until you hear your name, the name of your department, or some specific information that is directly relevant to your concerns. You will be a far more effective negotiator if you listen to the entire message.

Practicing good listening skills can help you with all these types of bad listening habits. Repetition of what you have heard helps you listen better, and avoid these bad habits. This is discussed in more detail later on.

Effective communication begins with patience and listening.

So, what do patience and listening get you?

CREDIBILITY

Patience and listening get you *information.* Information gives you knowledge. Knowledge leads to credibility—if you use it properly. The more you understand what the other party is looking for, the more credible you will be when you have something to say.

Chapter 3: Effective Communication

Credibility is the cornerstone of communication. You build credibility in order to be heard. It works like this:

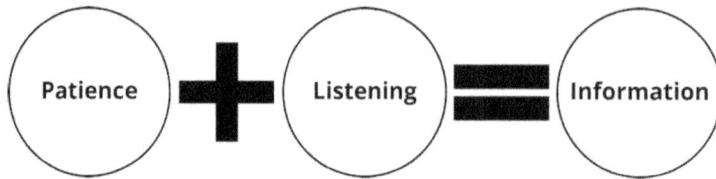

Figure 2: Patience + Listening = Information

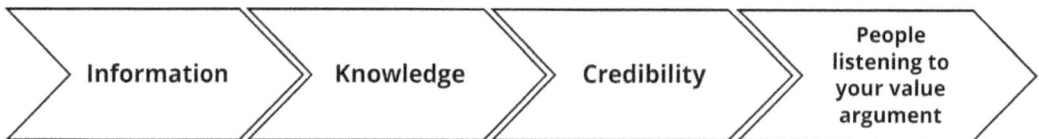

Figure 3: The Credibility Equation

Credible speakers:

- Listen carefully

- Don't promise more than they can deliver

- Treat people as individuals, not stereotypes, and have the facts

- Do not reject out of hand what people say

- Anticipate the effects of their speech

- Accept responsibility for what they say

When you have credibility, you can inspire belief. Often, we admire people whom we find credible. They inspire trust; they may even become personal heroes.

Whom do you find credible? In the workbook, list five people you think have credibility. Next to their names, list the qualities that make them credible.

Chapter 3: Effective Communication

STOP if you are using the Companion Workbook.

Exercise 3-3: Credible people

Credibility is the power to inspire belief.

Fortunately, people like you don't have trouble being credible. Maybe that's because you can point to what you know. And when you are credible, people tend to listen to you. This is critical in negotiations. Think about what happens when you are speaking with someone who has no credibility—someone you consider a "bullshit" artist. What do you do? You usually turn off; you stop listening. Well, if you are not listening to someone, what chance do they have of persuading you? However, just because you are credible does not mean you will be persuasive.

For example, you can easily present a business case showing why the price you're charging is competitive in your industry and is consistent with reasonable cost structures. However, to show the other side *why* they should pay that price—that it will make them more competitive in their business—you must know something *about* their business. That's more difficult, because you don't know as much about their business as you do about your own.

We can say that truly effective persuasive communication has three main attributes. It must be:

1. Understood
2. Believed
3. Valued

Let's look at these three qualities more closely:

1. Communication that is clear has a message that is *understood*.

2. Communication that is credible is *believed*.

3. Communication that conveys an awareness of your customer's business and needs is *valued*. It shows the benefit of what you do for them.

People don't make decisions because they understand or believe what you are saying. While these first two attributes are important, the main reason someone decides in your favor is the value you provide to them. Dealing with you makes them "better," however "better" is defined. Your value argument will usually have more impact if you can quantify "better".

The following chart shows the communication and negotiation process the K&R way. This is the K&R Leverage Cycle™.

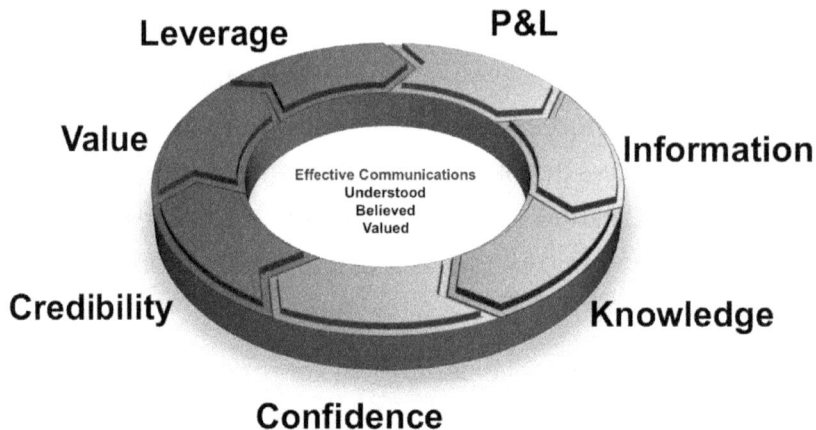

Figure 4: Leverage Cycle™

Chapter 3: Effective Communication

WHAT YOU LEARNED IN THIS CHAPTER

- **Communication** is the process of creating meaning. Communication arises from context.

- Our interpretation of words in a specific situation and the way we respond to them depend on our culture, experiences, values and emotions.

- Faulty logic can damage your credibility and the most carefully constructed value argument.

- Effective communication comes from P&L: Patience and Listening.

- There are three main kinds of listening: empathic listening, informational listening, and evaluative listening.

- Practice and repetition can cure poor listening habits.

- The most common poor listening habits are pseudo listening, self-centered listening and selective listening.

- Credibility is the cornerstone of communication. You build credibility in order to be heard. It works like this:

  Listening > Knowledge > Credibility

- Truly persuasive communication must be understandable, believable, and have value to the other side.

CHAPTER 4: CREDIBILITY AND LEVERAGE

Consider the following scenario:

M4 Software had a poor relationship with the Sentinel Account. The problem involved $10 million in previously purchased software that Sentinel had not yet installed. This "shelfware" had seriously strained the relationship. M4 sent in a new software account manager and new salespeople, but still couldn't sell value to Sentinel because they had a credibility problem. How did the M4 team deal with the trust issue? They used effective communication. They asked lots and lots of questions; they listened carefully. Sure enough, they uncovered the real problem.

The M4 team found an education gap. Sentinel had so much shelfware because they did not know how to use what they had already purchased. M4 solved the problem by educating the people at Sentinel. In this situation, M4 elected to absorb training costs to help the customer use the software they had already purchased. Trust was regained and a relationship was reestablished. Not long after, M4 closed a sizable new sale to Sentinel.

The K&R Deal Forensic

Here is what worked:

1. Resolving the strained relationship required knowledge of the causes of that strain.
2. When knowledge and understanding of the customer problem were obtained, M4 was able to create value that solved the cause of the problem (providing training to enable the customer to use the software).
3. Creating a solution to the customer problem earned the necessary credibility to build a relationship for the next deal.

This story shows how credibility creates the opportunity that allows leverage to shift. Read on to learn more about the intertwining of these crucial negotiating concepts.

WHAT LEVERAGE IS

What is **leverage**? Think about what you learned in high school physics: levers relate to movement, those clever pulley systems. Levers give you the power to accomplish seemingly impossible tasks. "Give me a lever long enough and a place to stand," Archimedes wrote, "and I will move the world." The longer the lever or the more levers in a pulley system, the more easily we can move objects of heavier mass using less force.

In a negotiation context, leverage helps you persuade people to move closer to your way of thinking. M4 used trust and credibility to help them regain leverage that resulted in a sale. Credibility creates the atmosphere for leverage by getting the other side to listen. Then you have the opportunity to articulate value arguments and have them heard and believed. That creates leverage.

The process looks like this (just like the Leverage Cycle™):

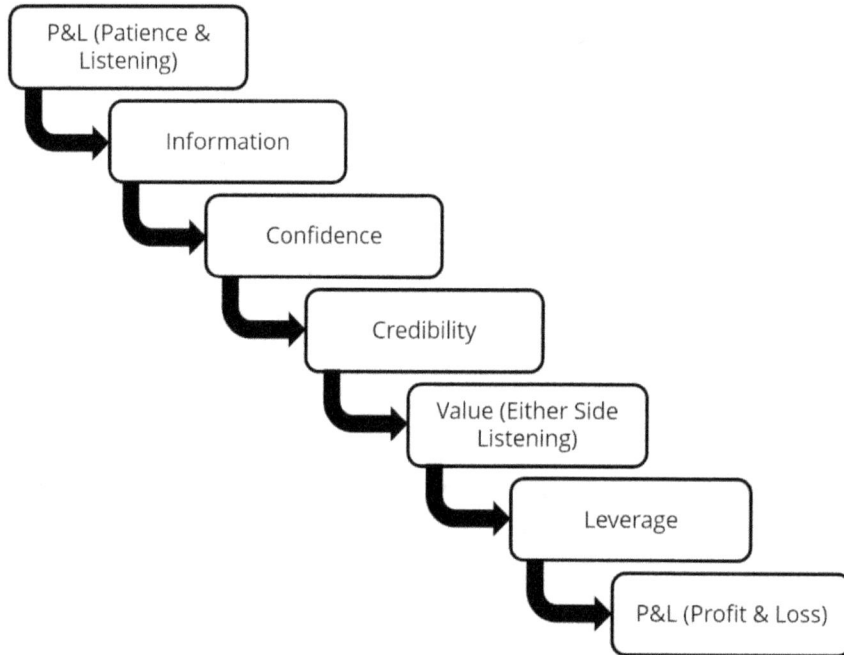

Figure 1: The Credibility - Leverage Process

The person with leverage has the advantage in a negotiation. So how do you get your audience to move closer to your way of thinking? How do you get them to pay the higher price rather than the lower price? How do you get leverage? One of the best ways is to quantify the positive impact of a favorable decision to them.

Leverage is the ability to persuade people to move closer to your way of thinking.

Use the workbook to describe a situation when you used leverage to move the other side closer to your position. Perhaps you used leverage at the negotiation table at work or while buying a large consumer item. It's just as likely you used it on the home front, dealing with your significant other or your children!

STOP if you are using the Companion Workbook.

Exercise 4-1: Leverage in action

POSITIVE VERSUS NEGATIVE LEVERAGE

Consider this scenario:

> *Today, Vendor is the only widget seller, although a year from now it is expected that there will be many other widget sellers. The customer needs widgets this week.*
>
> *How is the customer likely to deal with Vendor a year from now if situation #1 occurs now?*
>
> *Situation #1: Vendor says, "This is the price. If you don't like it, go somewhere else."*
>
> *How is the customer likely to deal with Vendor a year from now if situation #2 occurs now?*
>
> *Situation #2: Vendor says, "I appreciate your concerns about price. Please understand that we are supply-constrained. While we will treat you as fairly as we treat our other customers, we may not be able to do anything about the price at this time. Of course, we can revisit these issues in the future."*

Of course, you're likely to get much better results with situation #2. Allowing the other side to "save face," to avoid humiliation, does not mean giving up your leverage. Rather, it helps cement the business relationship. This is good for business. Forcing people to look for alternatives because of arrogance born of misused leverage is bad for business. You'll learn more about "saving face" in Chapter 12.

Unfortunately, the concept of leverage is often grossly misunderstood. That's because leverage gets viewed negatively, as a hammer used to bludgeon the other side into submission. While being the only game in town is a form of leverage, the best way to approach the situation is to have a position where the other side comes to your way of thinking voluntarily.

The Godfather, Mario Puzo's fictional saga of the mob, has become part of popular culture. Godfather Vito Corleone is famous for his blunt way of "negotiating": "Make him an offer he can't refuse," he said. Loosely translated, this means you either do what the Godfather wants or you find your horse's head between your silk sheets. This isn't using positive leverage. This is using fear and intimidation, a form of negative leverage.

Here are a few more examples of a win-lose style applied when one side feels it has leverage:

> *"Take it or leave it!"*
>
> *"Here's the deal: all or nothing."*
>
> *"It's my way or the highway!"*
>
> *"Take what we offered or we walk."*
>
> *...and the famous family ultimatum: "Eat it or go to bed hungry!"*

If Joe's solution is the only game in town, how does the other side expect Joe to behave even before they sit down to negotiate? Usually, the first words that we hear when we ask that question are "arrogant", "inflexible" and "obnoxious". When someone acts arrogant, inflexible or obnoxious with us, what is our natural reaction? We push back, become difficult, or search for whatever other alternatives there are, even if they are inferior. No one likes to feel coerced.

This is a natural "fight or flight" response. For example, a few years back, did a number of customers move away from Microsoft operating systems or IBM mainframes because alternatives were superior or because these companies were at times perceived as arrogant and difficult to do business with?

STOP if you are using the Companion Workbook.

Exercise 4-2: If I had a hammer

Read both of the following scenarios. Decide which, if either, use leverage wisely. Mark your answer in the workbook.

Scenario 1

> During a negotiation, the V.P. of "Z" side says to "G", What's all the fuss? We have a jump on the market; we're the only game in town. Everyone knows it. Here's the offer we will be making. For the time being, you can take it or leave it. G made the deal with Z, but made arrangements to switch as soon as an alternative was available.

Scenario 2

> Not too long ago, we were in negotiations on behalf of CCC with an important supplier; we'll call them Boris and Natasha (B&N, for short).
>
> The negotiations had reached an impasse because B&N was unwilling to disclose the next generation technology to CCC, who was also an B&N competitor. B&N's current product line was falling in the marketplace. CCC's adoption of B&N's future technology would likely catapult both B&N's current and future product line to success. CCC's marketing team that wanted to receive the information presented a persuasive business case showing B&N the advantages of disclosing the technology. They used their knowledge of both sides to fashion a convincing case. They were credible. B&N changed its mind, and a

disclosure was made with the right restrictions. This fostered a healthy relationship, then and for the future.

The discussion below adresses the issues raised in Exercise 4-2: If I had a hammer.

The K&R Deal Forensic

In Scenario 1, leverage was not used wisely.
1. Leverage from lack of alternatives was exploited so G felt coerced.
2. People don't like to be pushed around.
3. G's reaction to look for an alternative was predictable. This is an example of what happens when leverage is used as a hammer.

Leverage is the ability to move people closer to your way of thinking. Of course, leverage can come in the form of a hammer...but when it does, it can damage relationships. Often, successful leverage comes as a result of effective persuasive techniques, which are based on credibility. That's why the two concepts are so closely intertwined.

The K&R Deal Forensic

In Scenario 2, leverage was created through persuasion, facts, and logic.
1. CCC got the supplier to move closer to its way of thinking by persuading them with CCC's reasoning.
2. CCC used logic to persuade rather than strong-arm tactics.
3. This is using leverage positively. This helps build relationships rather than damage them. Interestingly, this type of approach was available to the players in Scenario 1 as well.

When people feel like they are being victimized (especially by intimidation), they will try very hard to find alternatives—even inferior ones—to doing business with you. If they have no adequate alternatives, they may do business now, but will switch to another partner or supplier as soon as the opportunity presents itself.

RECOGNIZING LEVERAGE

Consider the following story.

> *A man has a parrot that could swear like a sailor. The parrot could swear for five minutes straight without repeating himself. The problem is, the man who owns the parrot is a quiet, conservative type and the bird's foul mouth is driving him crazy.*
>
> *One day, the parrot's blue streak gets to be too much for its owner, who grabs the bird by the throat, shakes it really hard, and yells, "Stop it!" This just infuriates the bird and it swears more colorfully than ever. The man says, "OK, you; this is it!" and locks the bird in a kitchen cabinet. This infuriates the bird even more so it claws and scratches the wood. When the man finally releases the bird, our fowl friend cuts loose with a stream of vulgarities that would make a late-night comedian blush.*
>
> *At that point, the man is so angry that he throws the bird into the freezer. For the first few seconds there is a terrible din. The parrot kicks, claws, and thrashes. Then the room suddenly gets very quiet. At first the man just waits, but soon he gets worried. What if the parrot is hurt?*
>
> *After a few minutes of silence, the man is so worried that he opens up the freezer door. The parrot calmly climbs onto the man's arm and says, "Awfully sorry about the trouble I gave you. I'll do my best to improve my vocabulary from now on."*
>
> *The man is astonished at the bird's transformation. Then the parrot says, "By the way, what did the chicken do?"*

Especially as a seller, you are in a relationship business. Treat people well or they will take their business elsewhere. They could also try to get even. No one wants to feel like the man with the foul-mouthed parrot.

The following diagram shows the relationship of price and value:

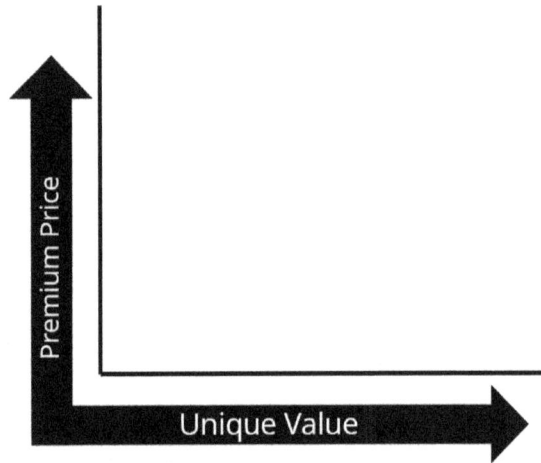

Figure 2: Recognizing Leverage - 1

Let's see how that relationship changes between a commodity-like and a monopoly-like situation in a transaction:

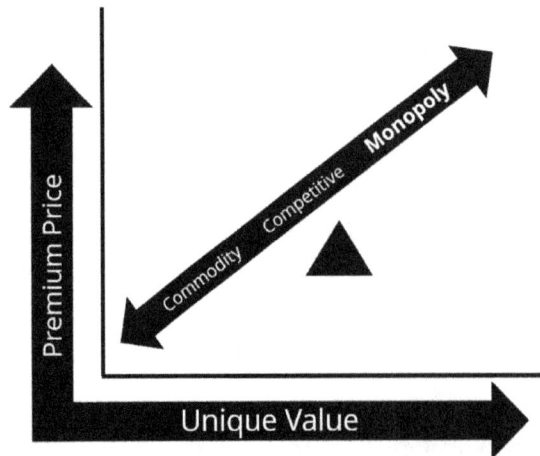

Figure 3: Recognizing Leverage - 2

Commodity: All offerings look the same and there are multiple alternatives; consequently, price is low.

Competitive: More than one acceptable solution, but each vendor has some unique value that can command a higher price.

Monopoly: Your offering is clearly the best customer choice for that transaction, commanding the highest price.

- **Procurement's job:** Most procurement acquisitions are measured by cost savings. The job of the buyer in a negotiation is to get all items to look the same, like commodities. This reduces product solution uniqueness and your chance to gain a monopoly position for that transaction. It results in lower prices.

- **Seller's job:** The job of the seller in a negotiation is to get as high up the slope as possible. The higher up the slope you can go, the higher the unique value of your product or solution. Thus, you can command a higher price. To get to the top of the slope, you need to establish your unique value, preferably quantified.

So, what's the best solution to a customer problem? The one that delivers a unique, quantifiable business impact for the customer: unique value!

COMMODITIES

In addressing the charts on the previous pages, let's first consider commodity products or services. These types of products or services are commonly available from many vendors. As you would expect, they do not have much differentiation or unique value: they look the same, tend to function the same, and are easy to procure. As a result, the key purchasing criteria tends to be price. The competition for commodity offerings tends to drive price lower and lower.

Chapter 4: Credibility and Leverage

Now factor in procurement's mission. In most companies, procurement's mission is to fulfill its company's needs at the lowest possible cost. Therefore, it is in procurement's best interest to try to convince the seller that their product or service is a commodity. By definition, that would mean you are one of many alternatives driving prices down. So be careful not to get trapped into a price-only discussion. Often the price comparison most favorable to procurement's argument ignores indirect cost savings that would result in a lower total cost of ownership. The seller's job—or your job if you are the seller—is the exact opposite. The seller's job is to make a value argument and convince the buyer that the seller's product or service is better than the competitors' in a way that impacts the buyer, thereby justifying a higher price. The following graphic shows this relationship in visual form.

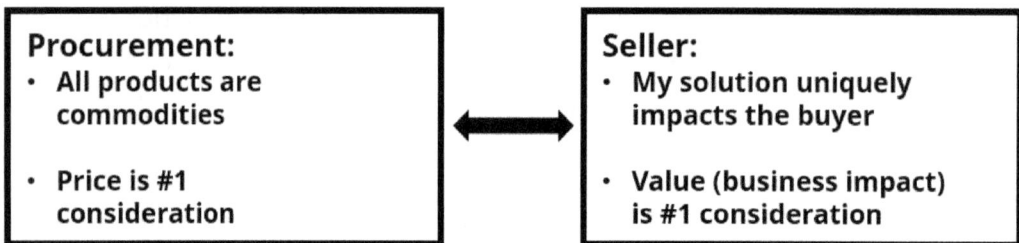

Procurement:	Seller:
• All products are commodities	• My solution uniquely impacts the buyer
• Price is #1 consideration	• Value (business impact) is #1 consideration

Figure 4: Procurement - Seller Relationship

COMPETITIVE PRODUCTS OR SERVICES

Competitive products or services have some special value that separate the seller's offering from most, but not all, competitive offerings. The seller can articulate some unique value as a competitive advantage for the buyer, enabling the seller to charge higher prices. This is usually an environment where more than one vendor has some uniqueness. Think price elasticity of demand in economics.

Chapter 4: Credibility and Leverage

Demand becomes relatively inelastic if the seller does a good job articulating the impact of its product/service on the buyer; thus, the buyer is willing to pay a higher price. However, there is a price point at which the buyer will purchase alternative solutions. These solutions may not be as effective, yet still achieve the buyer's overall goals. If you are a seller who is able to convince the buyer of some uniqueness, you will be in a better position to charge a higher price than you would with a commodity offering.

TRANSACTIONAL MONOPOLY POSITION

Consider this pun:

> In the late 1800s, the man who was shot out of the cannon every day at the Barnum and Bailey Circus decided to quit because his wife had asked him to find a less risky way of making a living. P.T. Barnum hated to lose a good man, so he sent him a message, "I beg you to reconsider—men of your caliber are hard to find."

What's your best position as a seller? You want to be the "man or woman of caliber". The best position to have in a transaction is to offer the best choice for the achievement of the other side's goals. When the seller is in a monopoly position in a transaction with unique, quantified value, the customer has no practical alternatives. If doing nothing is not an option, then demand becomes very inelastic. As a result, the seller can command a relatively high price. Of course, a smart seller still uses leverage wisely – as positive and not negative leverage.

A high-leverage position comes as a result of effective demonstration of unique value, and it is best when quantified in a credible manner.

KEY POINTS: RECOGNIZING LEVERAGE

- When you have the leverage of a monopoly position for a transaction, sustain your price. Do not move prices down the slope. Remember: When you are in a commodity position, the buyer will not pay high prices. Thus, to have a balance over a series of transactions, you must use those opportunities when you have a monopoly-like value, to sustain a higher average revenue stream.

- When you are in a monopoly position, use your leverage wisely. Explain the value to ensure the buyer feels they are getting the value. The customer should always feel good about the decision and the relationship you have forged between buyer and seller. This will benefit the seller in a subsequent transaction when they are not in a monopoly-like position. And it benefits the buyer, who gets the value and doesn't feel like they must search for alternatives to arrogant, monopolistic sellers.

MAKE A VALUE ARGUMENT

How do you move a client to close a deal? You communicate, you listen, and you establish value. To do so, make a **value argument**.

A value argument is a persuasive argument. In a negotiation, credible value arguments earn positive leverage.

Value in a deal comes from advantage given to both sides. For example, why would Company A and Company B want to do business together? The reason is simple: mutual benefits. Whether it is software or drilling equipment, each company wants something the other can provide. They will not discuss a relationship unless it provides business value.

Chapter 4: Credibility and Leverage

Delivering value to the other side is one of the easiest ways to get them to agree with you. If they perceive that you are offering something that impacts their business that other alternatives do not, they are more likely to make the deal.

This is especially important because no one likes to be victimized. If people feel cheated, they may try to get even.

Over years of negotiating deals as well as teaching people to negotiate in a business environment, we've discovered that people have trouble articulating and quantifying value (impact) to the other side. In fact, it's one of the single biggest problems we see!

That's why you need to do your homework! You're a smart person. You know to gather information and compare it before you rush into a negotiation. In Chapters 8 and 14, we will cover specific techniques to help you prepare for a negotiation.

BUILD A VALUE CASE

If you don't build a value case for the product or service you're selling, the other side may not see that value. Even if they do see the value, they may not acknowledge it, since acknowledgement of that value gives you leverage.

Additionally, not articulating value may affect your credibility. The buyer may feel the seller is not listening to what matters to them. As a result, the seller loses credibility.

We have observed that most people take their own value and leverage for granted. When working on a deal, we always ask our clients, "Tell us about your leverage." Often we hear, "We have no leverage; we are just like everyone else." So how do we turn around this defeatist attitude? We ask questions to unravel the leverage problem. Here is a sample conversation:

Question: *"Are you bigger than your competitors?"*

Answer: *"Some or most of them."*

Question: *"Have you been around for a long time?"*

Answer: *"Relatively, yes."*

Question: *"Can size or longevity provide assurance to a customer?"*

Answer: *"Of course."*

Question: *"What is your industry reputation as a supplier?"*

Answer: *"We were picked by* Procurement Monthly *as one of the most reliable suppliers for the past three years."*

Question: *"Do your customers rely on you for service and support?"*

Answer: *"Yes."*

Question: *"Well, then, you have a positive history with that customer and others, don't you?"*

Answer: *"Yes, but they understand all that."*

Now you have some information to begin formulating value arguments that will translate to leverage. For example, what if your capabilities as a supplier had previously resulted in incremental revenues or cost savings for your customer?

That would give you a credible history with this customer. Doing business again should be easier than starting from the unknown. Alternatively, a positive track record with other similar customers can support a positive business case for this one. It gives you an advantage, provided that you've performed. Or, what if your potential customer had done business with a younger, smaller company that is no longer in business? Perhaps longevity becomes a key factor in determining future business partners.

Unfortunately, most of us tend to take our strengths in negotiations for granted. We assume the other side knows the value we bring to them.

Remember, the customer does not owe you the recognition of your credibility, value or leverage. It's up to you to make the argument!

Articulation of value requires you to know something about the other side. The more you know, the better. Knowledge is power—especially in a negotiation. You have to know the business gaps they have to fill—and fill them. When you use value properly, you are usually successful.

You say to your customer, "When you buy this product, you'll get better response time, better delivery, and more satisfied customers." However, while this kind of vague statement may be believed and understood—and may even be persuasive to your advocate in the customer's organization—it usually isn't enough to induce a buying decision on the part of the customer. That's because your advocate or sponsor needs more of a business case to present to their company to get approval for the purchase decision. That's why quantifying value in the context of a customer business is so important.

Let's look at an example:

> We were asked to help a team negotiating with a worldwide travel agency. Our client had done a benchmark study for the customer that showed that their solution was 35% faster than both the competition and the current method the customer was using for processing transactions. In the negotiations, our client's sales team thought this 35% faster benchmark was self-explanatory, since the benchmark was witnessed by customer personnel. Our client wanted a price of $3 million to license and service this solution for the customer. It was the beginning of December and the customer countered with $550,000.
>
> The usual tendency in these situations, especially with the end-of-year pressure for the seller, would be to try to move much closer to the customer's position. That means reducing price to get whatever revenue

we could before year-end. However, we advised our client that it would be foolish to reduce price as long as the value to the customer merits the higher price. Instead, we asked the client to quantify the value. They did. Based on their knowledge of what the customer would save and additional revenue the customer would gain with the 35% performance improvement, that value justified a price as high as $3.75 million and provided a solid annual return on the customer investment (ROI).

So, the team went back to the customer with a business case that justified a $3.75 million price. That's right, we advised our client to show the customer a price of $3.75 million. The evidence was irrefutable and the customer quickly understood that the original $3 million price was a great deal. To allay the customer's concern that the price would now increase, our team said they would honor the earlier $3 million offer, but that offer would be off the table after December 31. The deal got done for $3 million before December 31! And the customer felt great paying "only" $3 million.

The K&R Deal Forensic

This transaction reflects several very good negotiation practices.

1. The team was open to new ideas and utilized excellent teamwork.
2. The team got back to value. The best relationships are built with value, not with discounting.
3. Leverage was created by the clear quantification of value, using data known to the customer.
4. That leverage was wisely used to get closure and build a relationship with a satisfied customer.

When you establish and quantify value credibly and move the transaction to a monopoly-like position (i.e., unique value), your customer will want to buy your product or service. You get leverage by working with the customer to show what your unique value means to their company.

The final and most difficult part of value is quantifying it. Quantifying value means taking it above articulating qualitative benefits to the customer and expressing those benefits either financially or in some other manner that matches a measurable customer goal. Even though quantifying value is often the most difficult part of the process, it is also the most persuasive aspect. More on this later.

PERSONAL VALUE VERSUS BUSINESS VALUE

Some of the best value you can articulate is the kind that makes the job easier for your company and for the other side. When you present them with a positive business case that shortens the workload they need to get internal approval — that makes their job easier. And that's when they are more likely to become a better advocate for you! But remember that the business case still must contain company value, which should be expressed as a quantified benefit (e.g., revenue, profit, cost savings, resources, efficiency…) to the customer.

Higher profit and lower loss come from a well-articulated value statement. Effective value statements show that you understand what your customer is looking for. Then customers will want to make the buy decision because that decision helps them. (And we'd all rather have a "good buy" than a "good bye"!)

Persuasive communication requires value. If you don't understand how the product or service achieves value for the customer, how can you articulate and quantify value to that customer? And your customer will be less likely to buy. Customers prefer to know the benefits they are getting from an investment. Remember they have to get approval based on a business case, so talk your customer's language. That's how you will be persuasive.

Here are some examples. **Don't** talk this way:

Poor Communication

- "Buy my service."

- "Buy my service, please."

- "Buy my product; it can help you."

- "Buy my product; I need this deal so I can make quota."

All this person is doing is projecting, not giving you a reason to listen or to buy the product. They are not speaking the customer's language.

Instead, **do** talk this way:

Credible Communication

Credible communication is a step to persuasive communication, but may not itself be persuasive.

- "Our product has these functions and here is the documentation."

- "We have seen the benchmark in your environment and our solution performs 35% faster. Here is the data."

By itself, credible communication gets the other side listening and even believing, but it generally does not indicate the benefits that are required to motivate a decision.

You also need to speak this way:

Persuasive Communication

- "If you invest in this solution, it will enhance your customer satisfaction, which means your sales will rise. Independent industry studies (here they are) show that for each 2% improvement in customer satisfaction, sales improve by 5%. Please feel free to verify them and let us know what you think. Based on your own data, this means X million in sales and Y million in profit to your company."

- "Our product will allow for faster processing, which means that you will save X amount over three years. These are the calculations we used. Please review them and let's move forward to implementation."

- "If you invest in our solution, you will not have to hire those three extra salespeople, so you'll save an estimated X amount of money, improve your E/R (expense/revenue ratio) and, as a bonus, have more loyalty, less turnover, and reduced training costs. Here is the data that supports this decision."

- "Here is the data and some references from similar situations we have been involved with. Note the ROI, IRR, and improved market share that helps solve the problems you and I have discussed. The value to you is that you will reduce your operating costs, enhance customer satisfaction, and provide better service."

Through credible communication, you give the customer a simple reason to listen. When you articulate and quantify value, you become persuasive so they will buy your product or service. Remember: Most people are more interested in themselves than in helping you achieve your objectives (unless it's your family member).

Try it yourself. Think of a market for your product or service. Then list three reasons why the customer should buy the product or service. Think of how each reason can be tied to some uniqueness of what you are offering. Make sure each reason matches their needs—not just yours.

Think about how these reasons could be quantified.

To get you started, here are some examples of customer concerns—reasons for them to consider your product—in a retail environment:

> **Reason #1:** *"Does this drive business into my retail environment? What volume of business does this affect?"*
>
> **Reason #2:** *"Can I answer customers' inquiries faster with this product? Does this enable me to reduce resources? Recognize revenue faster?"*
>
> **Reason #3:** *"Can I provide faster access online with this product? What does this mean in terms of my productivity?"*

In the workbook, list the market you serve, then articulate value for your products or services.

STOP if you are using the Companion Workbook.

Exercise 4-3: Articulating value

Let's pull it together.

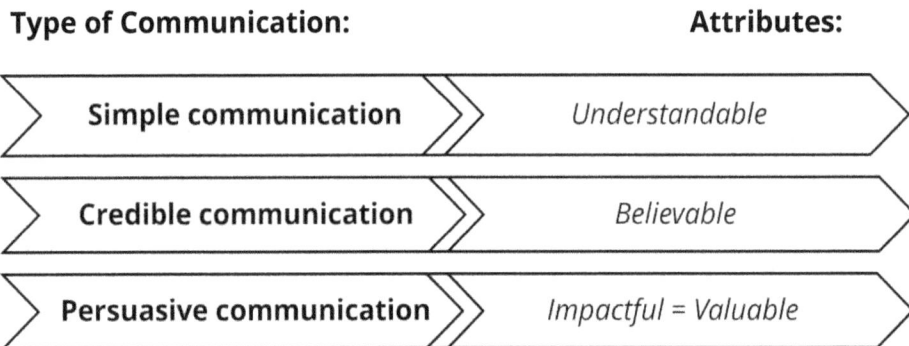

Type of Communication: **Attributes:**

Simple communication	Understandable
Credible communication	Believable
Persuasive communication	Impactful = Valuable

Figure 5: Communication

Chapter 4: Credibility and Leverage

WHAT YOU LEARNED IN THIS CHAPTER

- Leverage is the ability to persuade people to move closer to your position. The best leverage is not intimidation. It is positive leverage from a perception of unique value that impacts your customer so they willingly agree to your terms and prices.

- A value argument is a persuasive argument. In a negotiation, credible value arguments lead to leverage.

- You must build a business case for the solution you're selling that shows a quantified impact of that solution for the customer.

- The customer does not owe you recognition of your value; *you* must make the argument!

- Articulation of value requires you to know something about the other side. The more you know, the better.

- Effective persuasive communication must contain three attributes: it must be clear (understandable), it must be believable, and it must contain value for the other side.

- Procurement often has a cost measurement; give them a good business rationale to help their business overcome the cost argument and decide in your favor.

CHAPTER 5: CREDIBILITY AND LEVERAGE—THE FOUR SCENARIOS

I had just been promoted and was told I was the youngest branch manager in the company. As he promoted me, my boss said, "You're going to make a lot of mistakes. Just try to be creative enough to make a different mistake every time." I'm sure I've exceeded his expectations!

—*HARVEY*

The boss was not exactly giving Harvey a vote of confidence, but his advice did motivate Harvey to try more creative solutions.

It is now time to use what you've learned so far about the art and science of negotiation, creativity, effective communication, credibility and leverage. We are going to present four real-life scenarios that all require the art and science of negotiation. Each scenario represents a challenging situation dealing with credibility and leverage.

Most people are prone to reacting precipitously when confronted with a difficult question or situation. But, it is precisely at such moments that thoughtful responses and actions are most critical to maintaining credibility and leverage. When confronted with a tough situation, you'll often find that either credibility or leverage is at stake—usually both. When these situations occur, be thoughtful in your response. This will help you minimize any negative effect. Your carefully thought-out, reasoned response can also help improve your overall position.

However, it is easy to say, "Be thoughtful and reply with a reasoned response." Doing it when there's emotion about the stakes involved is a different matter. So how do we do it?

Chapter 5: Credibility and Leverage—The Four Scenarios

One of the best ways is to work with a team, even if it means just finding a trusted colleague, to help you come up with a logical solution. If you work with a team, the chances of having someone who steps back to bring reason to an emotional situation increases. Your teammate will get you back on track. As you debate the issues and work out the best responses, consider what we've discussed about P&L (patience and listening), credibility, value, and leverage.

As you work out your negotiation strategy, consider three key questions:

- What is at stake in the situation?

- What problem am I trying to solve?

- How can I articulate and quantify value?

After you have developed your negotiation strategies, write them down in the workbook and compare them to ours. Don't forget: Very rarely in negotiations is there only one way to approach—and solve—a problem. There are usually several alternatives to achieve a goal.

Remember this K&R Axiom: Each action taken in negotiation can affect credibility or leverage in some way...so be aware.

STOP if you are using the Companion Workbook.

Exercise 5-1 through 5-4: MEDAPP, Misguided integrity, Go, go CEO, Fail test

Scenario 1: MEDAPP

You represent MEDAPP, a company that sells and manages industry-leading medical/healthcare applications, for many large hospital facilities. Today, in about 90 minutes, you and your sales team are meeting with a client, with the goal of selling your software package to them.

During the team's breakfast meeting, you glance at a news report and see a headline and article that states your company's healthcare application, installed at another hospital, has crashed and may be involved with the deaths of two patients and the critical condition of 17 others. In 90 minutes, you are supposed to be presenting the same solution that is in this newspaper article to your client.

Scenario 2: Misguided integrity

Many believe being "open and honest" in a negotiation is the best approach. You are meeting next week with a manufacturing company that is going to produce chips for your line of mobile devices.

Last week, you dismissed your alternate manufacturer and are now left with only this sole source. Your entire strategy is dependent on the manufacturer and you want them to know how important they are to your success. So, you plan to tell them the following: "Our entire strategy is dependent on you because we have no other alternatives. Please do the best you can for us. Thank you."

Scenario 3: Go, go, CEO

You are the CEO of a company in trouble. You need to finalize a deal with Big, Inc. in ten days and then present it the Board of Directors. This deal is important on several fronts, not the least of which is personal— this deal will allow you to keep your job. You try to make this a good deal for your company, of course. To that end, you have hired a consultant, a well-known and well-respected professional negotiator in the industry.

Now the deadline is four days away and some key issues remain open. The lead negotiator from Big walks in and tells you the following: "Your lead negotiator is doing a great job of raising all the right issues. However, since these issues take time to resolve, we are not going to meet the deadline for your Board of Directors meeting. I just wanted to let you know that."

Scenario 4: Fail test

Your new technology is in testing and is failing at a rate of 40%, which is ten times higher than your projected rates at general availability in sixty days (a date which is already announced). The testing data has been rumored in the market and you have actually seen similar numbers in industry publications. You are meeting today with an important customer who is sensitive to quality.

STRATEGY SESSION

By now, you realize that the four scenarios in the workbook represent challenging situations dealing with credibility and leverage.

As we mentioned at the beginning of this chapter, many people are prone to reacting prematurely and emotionally when confronted with difficult questions or situations such as these. You may know these people: the knee-jerk crowd that embraces the first strategy that comes to mind. Often, they cling to a strategy despite good advice to the contrary. That's not you, right?

When times get tough, you know that thoughtful responses and actions are most critical to maintaining credibility and leverage. Tough situations like these usually hinge on either credibility or leverage—and often both. To minimize the negative effects of what is at stake, be thoughtful in your response. Being calm, cool, and composed will probably improve your overall position. And what helps you be thoughtful, calm, and cool? Think "Be prepared!" Think "Teamwork!"

Here are some choices you may have made and the choices we made.

SCENARIO 1: MEDAPP

Below are some possible options and some of our favorites. See how these compare with your decisions.

- **Option 1:** Bail. You cancel the meeting, punt...and return calls to headhunters.

- **Option 2:** Call your key contacts at the home office, discuss the situation, and see how big the issue is. Consider whether the situation really is as serious as it seemed from the newspaper report. Get some advice, and maybe let your key contact at headquarters open the meeting (by speakerphone, if necessary) to defuse the situation.

- **Option 3:** Give your presentation as if nothing had occurred. Hope they don't raise the issue. If asked about the news report, pretend you didn't read it and say you have no information at this time.

- **Option 4:** Gather whatever information is available from your peers prior to the meeting. At the meeting, address the situation head-on. Show empathy and explain how well your company reacts to business disasters. Explain that you, your team, your support staff, and all of MEDAPP are in the process of gathering the information regarding this disaster and that you will keep them updated on what progress is being made. Assure them that, as a responsible member of the community, your company helps solve problems like this even if your product is not the cause; that yours is a good company that works with your customers to remedy bad situations.

Option 4 is probably the best choice.

What is *really* at stake here? What problem are you really trying to solve? Will you make a sale that day? Probably not. All you are trying to do is damage control to preserve an opportunity to make a sale. It is more than a big sale: It is future sales as well, because what's really on the line here is your reputation.

Your credibility is at stake. And not only **your** credibility, but also the credibility of your company, your customer, and your customer's company. The one thing you have the most control over is your own credibility. What does it say about your credibility if you leave or ignore the situation? Don't play dumb—what if your customer's people watch the same news reports? How are they likely to behave? They are likely to be hostile, attack you, and put you on the defensive. Or they may wait quietly, waiting to see if you bring it up. Either way, not a pretty picture.

But even if the customer didn't see the news report, think about their reaction after the meeting when their manager says, "Why are you dealing with MEDAPP? Didn't you read the news report?"

Now look at your options again. Why don't you want to cancel the meeting? If you do, wouldn't you lose your personal credibility and trust? Canceling the meeting would also make it appear that the problem is yours. Therefore, you must have the meeting and address the issue. The question is how.

Think about this strategy:

This is where art and science meet. First of all, before the meeting, do your best to gather the (presumably) little information that is available from your home office. Then, when you arrive, show them a copy of **the news report** and say, "Some of you may have seen this. May I tell you what I know and what I don't know? And let me tell you what a company like ours does about a problem like this even if we did **not** create the problem, because we are responsible citizens."

What starts happening? The customer will start listening because you are providing value to them. You are giving the customer a defense to offer their own executives who may raise the issue internally with questions such as:

"Why are you dealing with MEDAPP and that mess?" Remember the personal value we discussed earlier? That's what you are delivering.

Need further proof that we're right on target? Think about the Tylenol tampering situation. Johnson & Johnson pulled all the bottles from the shelves within 48 hours and ceased production until it could start manufacturing tamper-proof bottles. And they weren't even the ones doing the tampering! Because they faced the problem head-on, their market share in hospitals was unaffected. Thus, in a crisis, they enhanced their credibility and gained leverage in their market.

Contrast the Tylenol approach with many product recall situations such as Intel's, the Pentium® "Divide by Zero" fiasco:

- In June, Intel discovers a bug in the Pentium® processor. It takes months to change, re-verify, and put the corrected version into production. Intel plans good chips to be available in January the following year: four to five million units had been produced with the bug.

- In September, a scientist suspects an error and posts the suspicions on the Internet.

- On November 22, Intel issues a press release stating: "Can make errors in ninth digit…Most engineers and financial analysts need only four out of five digits. Theoretical mathematicians should be concerned…So far only heard from one." Intel claimed the error would happen once in 27,000 years for a typical spreadsheet user: 1,000 divides/day x error rate, assuming numbers are random.

- On December 12, IBM claims the error happens once per 24 days and bans Pentium® sales. Intel says that it regards IBM's decision to halt shipments of its Pentium® processor-based systems as unwarranted.

- On December 21, Intel takes a $500 million write-off. They also issue the following statement:

To owners of Pentium® processor-based computers and the PC community: We at Intel wish to sincerely apologize for our handling of the recently-publicized Pentium® processor flaw. The Intel Inside symbol means that your computer has a microprocessor second to none in quality and performance. Thousands of Intel employees work very hard to ensure that this is true. But no microprocessor is ever perfect. What Intel continues to believe is technically an extremely minor problem has taken on a life of its own. Although Intel firmly stands behind the quality of the current version of the Pentium® processor, we recognize that many users have concerns. We want to resolve these concerns. Intel will exchange the current version of the Pentium® processor for an updated version, in which the floating-point divide flaw is corrected, for any owner who requests it, free of charge any time during the life of their computer. Just call 1-800-628-8686.

Sincerely,

Andrew S. Grove	Craig R. Barrett	Gordon E. Moore
President/CEO	Executive Vice-President & COO	Chairman of the Board

---

As a result of having taken six months to acknowledge the problem, Intel wound up with a black eye that helped breathe life into its major competitor at that time.

The reason we say Option 4 is the best choice is: You acknowledge the problem from the customer view, even if it is not caused by you.

This straightforward approach preserves your credibility, gives reassurance to the customer's team, and allows you to keep a relationship sufficient to make a sale in the future.

SCENARIO 2: MISGUIDED INTEGRITY

It is true that sometimes you have to go with your instincts. And, sometimes the straightforward approach can be the best approach. But in this case, our view is that openness is "misguided" and unnecessarily damages your leverage. It hinders your ability to make a good deal. If you've severed contact with your alternatives, it doesn't mean you have to tell your vendor. There are other options that help you maintain leverage. Here are some of those possible options. See how they compare with your decisions.

- **Option 1:** If you already told them that they are your only option, go back to the alternative vendor you just dismissed, and restart discussions.

- **Option 2:** Even if you would rather deal with this seller, you have options. Tell the seller: "Look, we can go with you or your competitors. It's up to you!"

- **Option 3:** Make lemonade from the lemons you have, so next time say, "Of all our alternatives, you are our first choice. Let's see if we can strike a relationship so we won't need to explore those alternatives. But of course, that kind of close relationship would require your best terms and prices."

Once again, the last option is the best choice. The first option is probably good to the extent you reestablish contact with a potential supplier you may need in the future. However, that does not address your leverage problem in the short run. And going to the dismissed manufacturer quickly and apologetically may make them perceive that you have lost some leverage, making for a difficult renegotiation. Obviously, the second alternative leaves you no options. If the seller calls your bluff, you may be stuck without a solution.

The third alternative is an example of managing information. This allows you to offer an inducement of a "sole source" relationship for the vendor. The guarantee of exclusivity should induce the vendor to offer better pricing and terms. An additional advantage to you in this case could be reduced infrastructure (one product, one vendor), further reducing internal costs. However, the risk of going sole source is not a trivial one. If your one manufacturer has a supply problem, you have no alternatives.

Sole sourcing can often get you a better price because of higher volumes, time savings, and increased efficiency. Keep in mind the following: If you are a buyer, K&R does not endorse sole sourcing. If you elect to go sole source, it requires very careful contract construction that protects you and your organization, especially in the event of supply problems.

In conclusion, you should not have told the other side everything. Negative information needs to be managed, yet integrity needs to be maintained. Don't lie, but don't open your mouth unnecessarily. Later in the book, we'll talk more about how you can manage information while maintaining your credibility.

SCENARIO 3: GO, GO, CEO

Here are some possible options. See how they compare with your decisions.

- **Option 1:** Extend the deadline. A few more days won't matter. After all, what's a week, one way or the other, in the big picture?

- **Option 2:** Dismiss the negotiator and hire a new one—someone who can act fast. If all else fails, call off the negotiations.

- **Option 3:** Check that the deadline is real. If it is, you must act at once and get the deal done within that time frame. But you still want a good deal for your business.

Of these, Option 3 is probably the best choice.

Who had the pressure of a deadline here? It's the CEO on whom the deadline was imposed. A legitimate question is whether the deadline is real. This scenario is analogous to many we have seen. If it is a real deadline, who has the leverage if one side has that deadline? Usually it's the other side.

Why did the lead negotiator from Big walk in and tell the CEO (you), "Your lead negotiator is doing a great job raising all the right issues. However, since these issues take time to resolve, we are not going to meet the deadline for your Board of Directors meeting. I just wanted to let you know that." Big's lead negotiator wanted to get more favorable terms by creating pressure on the other side. This was a clever gambit for the negotiator; a potential problem for the CEO whose job was at stake.

There's an interesting dynamic in this scenario. As the CEO, right now you have the pressure because you want to keep your job, yet you do have some leverage if you can keep your emotions in check. Not all the leverage is with the other side. What happens if the deal is not completed by the deadline? Might Big have to start the negotiation all over again with a different CEO? If the deadline was breached, Big loses its leverage. This dynamic of shifting leverage is critical, especially in sales negotiations. Salespeople often give buyers additional incentives to close deals by certain end-of-quarter or end-of-year deadlines. Those incentives are intended to help compel or motivate the customer to act, so must expire by the deadline. Otherwise the leverage wouldn't shift away from the buyer. And then there would be no reason to offer the incentive in the first place. In fact, if you extend the reduced price beyond the deadlines you set, you will prolong the closing process, and impact both your credibility and your margins!

SCENARIO 4: FAIL TEST

Here are some possible options. See how these compare with your decisions.

- **Option 1**: Explain to the customer that in a test environment, errors are to be expected. You expect to find the bugs and fix them before the product is released into the field, and your good history with prior products reflects that. You may offer the customer an option to participate in the testing program.

- **Option 2**: Emphasize your superb support environment. The customer has nothing to worry about. You may offer some extra support if they commit now.

- **Option 3:** Evaluate and stress the value that the customer can derive from your product. Demonstrate the opportunity for the customer to get in on the ground floor with a valuable solution that will give them a competitive advantage. And try to remain firm on the price.

- **Option 4:** Offer a positive inducement for the risk the customer will be taking.

All four options have possibilities and can be used individually or presented as a complete response. After all, "it depends" is a key phrase in negotiation.

Your company has a credibility issue with the product. The buyer will have a problem with her own organization if she is looking at purchasing what is perceived as a flawed product. So you deal with the credibility issue by showing the buyer what her organization can get from committing to your product now.

And since the buyer is taking a risk, you might offer a reward. This sweetens the deal and compensates the buyer for the risk. Possible rewards include offering additional value, such as:

- **A discount:** Offer a specific discount if you fail to eliminate the bugs in a timely fashion. (But note, discounting itself does not address the risk issue!)

- **Priority service:** Offer priority service, as needed.

- **A test site:** The customer can test the product in their environment before availability so their issues are addressed.

- **Extended support:** Offer more favorable support terms over an extended period of time.

The customer provides a value to you by helping you reestablish credibility against the negative publicity. A smart customer will articulate that value as well as the risk they are taking by making a commitment to the seller now. But what happens if you resolve the quality problems before they decide to do business with you? The customer loses the leverage your credibility issue provides. The same thing happens if the seller strikes a deal prior to product availability with a different customer endorsing the product—the seller's credibility may be restored and the customer loses leverage. Similarly, as in Scenario 3: Go, Go, CEO, here, the customer's leverage is for a limited time; note how it shifts.

Whether you are the seller or the customer, your perception (based on industry knowledge) of all these factors, as well as the value of the product solution, is going to dictate what kind of offer you make. In making these offers, credibility and leverage work together to help you determine what will make business sense.

Axiom: Each action taken in negotiations can affect credibility or leverage in some way.

Chapter 5: Credibility and Leverage—The Four Scenarios

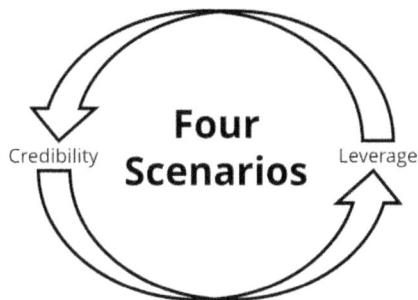

Figure 1: The K&R Negotiation Method™

One other thing these scenarios teach us:

It's always important to understand your leverage.

Don't be fooled: The elephant doesn't always have more leverage than the mouse. Small companies can have great leverage if they have cutting-edge technology, particularly if protected by strong intellectual property rights.

UNITED WE STAND, DIVIDED WE FALL: THE IMPORTANCE OF TEAMWORK

It's likely that you will complete these four exercises on your own.

Nonetheless, with respect to these four scenarios, keep in mind the value of teamwork. Wouldn't getting different perspectives from your team members help you come up with a better solution? Are you as thoughtful and successful if you do all these exercises by yourself? Teamwork in negotiations is crucial to success even in small companies where teams are small. You'll learn more about the importance of teamwork later.

WHAT YOU LEARNED IN THIS CHAPTER

- **Leverage** is the ability to persuade people to move closer to your position. Leverage is not intimidation; it is created by value.

- Use leverage wisely to build relationships, not ruin them.

- Articulation of value requires you to know something about the other side. The more you know, the better.

- Both credibility and leverage are essential to negotiating successfully.

- Credibility issues often lead to loss of leverage.

- Deadlines affect leverage.

- Leverage shifts in negotiation, so use the shifts wisely.

- The best "positive" leverage is a credible business case for the product or service you're selling.

- Before reacting, step back and ask: "What problem am I trying to solve?"

- Teamwork in negotiation is crucial to success. It helps us manage emotion and conduct ourselves logically with rationale.

CHAPTER 6: K&R'S SIX PRINCIPLES™

Speaking to the Young People's Society in Greenpoint, Brooklyn, in 1901, the famous writer Mark Twain advised: "Always do right. This will gratify some people and astonish the rest."

As you learned in Chapters 4 and 5, credibility leads to leverage. If you're credible, the customer is more likely to listen to you. If you don't have any credibility, the customer will more likely tune you out. In this chapter, you will discover K&R's Six Principles™, which govern successful negotiations. Following these principles will help you gain credibility and leverage in all your negotiations.

THE SIX PRINCIPLES

Ever hear the lyrics "More, more, more"? We all want more, whether we're the buyer or the seller. As the buyer, we want to get more for our money; as the seller, we want to earn more for the company and ourselves. The following Six Principles can help you get more, while staying true to yourself and ensuring good business for both sides.

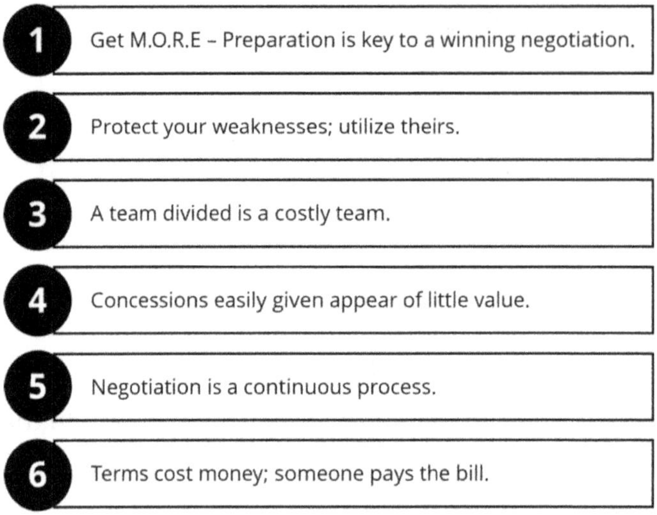

Figure 1: K&R's Six Principles™

PRINCIPLE #1: GET M.O.R.E.- PREPARATION IS KEY TO A WINNING NEGOTIATION

M.O.R.E. is an acronym, a word whose letters stand for other words. It's our easy way to help you remember the basics of successful negotiations. Here's what each letter stands for:

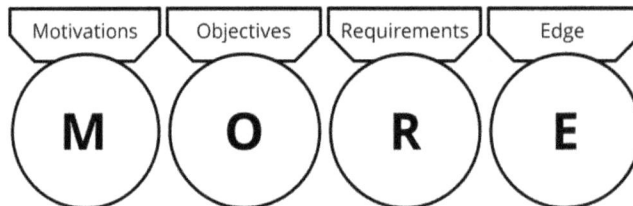

Figure 2: M.O.R.E.

Chapter 6: K&R's Six Principles™

Word	Definition	Explanation
Motivations	Motivations are the reasons someone acts. They answer the question "why?".	If you are a salesperson, your motivation is your quota that's been set by your management. As such, your motivations are behind you, driving your actions.
Objectives	Objectives are goals. They answer the question: "_What_ is someone trying to achieve?"	Two common objectives in business negotiations are revenue and profit. Since you seek these objectives, they are *ahead* of you, spurring you forward. If you are on quota, then the revenue or profit target is your objective.
Requirements	Are the conditions between you and your customer that will get you to your objectives and your customers to theirs? Requirements can also be defined as stipulations, qualifying factors, and terms that will be put into the contract that enable achievement of objectives and satisfaction of motivations. They answer the questions "How", "Who", and "When".	Requirements could include technical support, resources, maintenance, pricing, training, timing milestones and future upgrades. Requirements are between vendor and customer, but could include third parties.
Edge	Is the advantage you gain as a result of preparation and understanding M.O.R.E on both sides. Your "edge" gets the results you want when others have been turned aside; it's the additional leverage you have in a negotiation.	Having prepared, we gain confidence in our business case and have the backbone to stand by our value. So, we close on our terms and prices, while satisfying the other side.

PREPARATION IS KEY

Thorough preparation necessitates probing and gathering information. This helps you gain a more complete understanding of customer **motivations** that must shape the unique value-based solution from the seller. As a matter of practice, professional negotiators host planning sessions to gather the needed information to prepare for customer negotiation meetings. Information gathering, planning, and rehearsing will build a solid foundation of knowledge and confidence that will minimize (not eliminate) surprises when you are engaged in actual negotiations.

Motivations are characterized as circumstances—the whys—that have happened in the past which will dictate future buying behaviors and commitments. For example, a customer's position in the market, their service levels, personnel efficiencies, productivity, as well as their confidence in the supplier's products/services, capability, and reputation shape the need for and willingness to accept your proposal. Articulation of quantified value in the business case, expressed in terms that are relevant to the customer based on those motivations, is key to a winning proposal.

When you work harder and smarter in the preparation cycle, you gain knowledge and generate confidence. This improves momentum and increases leverage and credibility with the other side. After all, customers really want to do business with trusted suppliers who know their customer's problems and understand their business objectives. In other words, you need to know your customer's customer! Customers appreciate when suppliers are astute in the manner in which the solution is developed, articulated, and underpinned by a compelling business case. Take the time necessary to create unique value for your customers and be persuasive when presenting your solution. The rewards are worth the investment of time and planning. The benefit of preparation is that the time invested in planning early in the process saves time in closing the deal at the end. Of course, you have to judge when you've hit a point of diminishing returns on preparation given the size and importance of the deal.

IS MONEY THE SOLUTION?

We often see people "throw money at a problem" when money is not going to solve the real problem. Money is just perceived as a quick, easy fix. But how do you know if it addresses the real issue, the real goal?

Money is not always the requirement to meet your or the other side's objective. You must figure out the objective—what the problem really is—to develop the requirement that will get you there.

The following example illustrates this point.

> *We instinctively make assumptions that money or a lower price is a requirement when we might find out otherwise if we bothered to understand the customer's objective. Let's say that Roger Federer, the tennis great, retires and then changes his mind to make a comeback. He may be out of shape. That circumstance—being out of shape—is his* **motivation.** *As a result, his* **objective** *for a practice match with his friend Andy Murray this afternoon is to get into better shape. Given Roger's motivation and objective, where does he hit the ball? Most people would assume he hits it to the center of the court, so their rallies will be longer, and Roger will get a workout to get into better shape. Of course, just for fun, they keep score, and all of a sudden Andy is winning almost every point, as Roger is running all over the court. As Roger sees Andy smiling, while blasting another forehand past Roger, Roger's motivation begins to change – he is tired of losing points. Thus,*

his objective changes from wanting to get in shape to showing Murray who's boss. His objective becomes to win and he starts hitting the ball from sideline to sideline.

Without understanding the objectives, you are likely to be guessing what terms—the **requirements**—are mandatory for the customer. It is imperative that your internal planning and fact-finding activities for the customer be comprehensive. You try to leave nothing to chance. By planning early and often, by engaging a competent cross-functional team, and by exercising many questions and possibilities, you gain the **edge.** You will strengthen your odds of winning and close earlier with higher revenue, profit, and commission level. Remember: Everyone has **M.O.R...** the key is discovering and using them artfully to gain a winning **Edge.**

When you understand the motivations and objectives, you can fashion the requirements to meet these objectives. Sometimes the logical guess about the problem (the real objective) is the wrong one. We can think of numerous instances when price was lowered at the end of a quarter, but the deal didn't close anyway and the lowered price became the starting point for negotiations the next quarter. What type of concession is this: principled or unprincipled? We'll get to that shortly.

For our visual learners, M.O.R.E. looks like this:

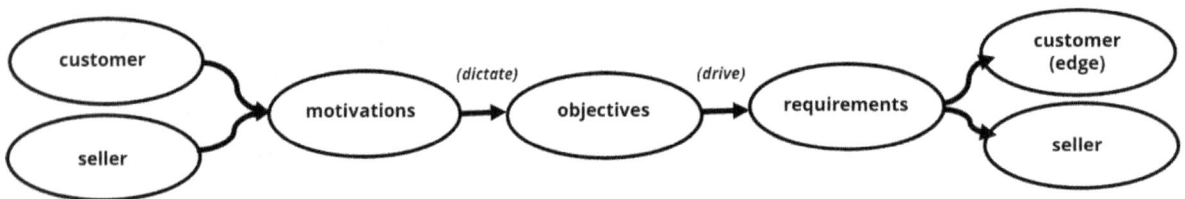

Figure 3: The M.O.R.E. Process

Chapter 6: K&R's Six Principles™

Ask yourself the following questions as you negotiate for M.O.R.E.:

Motivations

- Why would they be interested in doing business? (This question is about *them* – not *us*)

- Why would they be interested in doing business with *us?*

- Why would we be interested in doing business with *them?*

- Why are *we* doing this? (for each specific request)

- Why are *they* doing this? (for each specific request)

Objectives

- What are *we* seeking to achieve?

- What are *they* seeking to achieve?

- What do *we* want to get as a result of our relationship or as a result of this contract?

- What do *they* want to get as a result of our relationship or as a result of this contract?

Requirements

- How are *we* going to achieve our goals?

- How are *they* going to achieve their goals?

If you complete steps 1 (M.), 2 (O.), and 3 (R.) successfully, you will get 4 (E.), the Edge in a winning negotiation. Ultimately, this will lead to a naturally successful relationship with advantages to you = the Edge. And it will work for the other side as well!

In the workbook, list the three main **motivations** that drive your business, your top three business **objectives**, and the three most important business **requirements** that enable you to achieve your objectives.

STOP if you are using the Companion Workbook.

Exercise 6-1: M.O.R.E.

Keep these motivations, objectives, and requirements in mind as you read on. See how you can apply what you read in this chapter and throughout the book to help you get M.O.R.E.

Motivations, Objectives, Requirements, and **Edge** are the operational principles of a negotiation. Understanding how to use M.O.R.E. helps you grasp the ramifications of your actions and make conscious decisions that factor in how people are likely to react given their motivations and objectives.

PREPARATION IS KEY TO A WINNING NEGOTIATION.

Now, let's look at the second half of the first K&R Principle: *Preparation is key to a winning negotiation.* As you think about this sentence, ask yourself these questions:

- How do you go from hard work and knowledge to credibility?

- What actually happens? What is the process?

Preparation means always gathering information to gain an understanding of the motivations and objectives of the other side as well as our own. Without this understanding, we are merely guessing to figure out the correct terms (**requirements**) that might satisfy us and the other side. How can you solve the other side's problems if you don't know what they are?

BE PREPARED

The Boy Scouts were right: "Be Prepared" is a great motto. Often, success rests on preparation. When you go into a conference, speech, or any business meeting prepared, you exude confidence. The people in the room perceive that confidence and react with confidence in you, your product, and your expertise. Consider this little story:

> There was an engineer who had an exceptional gift for fixing all things mechanical. After serving his company loyally for over 30 years, he happily retired.

> Several years later, his company contacted him regarding a seemingly impossible problem they were having with one of their multimillion-dollar machines. They had tried everything and everyone to get the machine fixed, but to no avail. In desperation, they called on the retired engineer who had solved so many of their problems in the past.

> The engineer reluctantly took the challenge. He spent a day studying the huge machine. At the end of the day, he marked a small **x** in chalk on a particular component of the machine and proudly stated, "This is where your problem is." The part was replaced and the machine worked perfectly again.

> The company received a bill for $50,000 from the engineer for his services. They demanded an itemized accounting of his charges.

> The engineer responded briefly:

> One chalk mark: $1

> Knowing where to put it: $49,999

When you're prepared, you know where to put the chalk mark...and that makes all the difference. It doesn't matter if the "chalk mark" is the answer to an especially important question, a crucial sales point, or an issue of technical support. If you're prepared, you'll be better qualified to close the deal. Chapter 14 deals with preparation in detail.

Without adequate preparation, you lack the knowledge you need and you will lose confidence. And without confidence it's difficult to be believable, to have credibility. Even more important, with knowledge, you can prepare convincing arguments addressing issues the other side might raise in the negotiation process. That not only leads to credibility, but also to persuasion.

Consider the following two scenarios.

Scenario 1

Sandy Desmond and Pat Smith get a call from a client they know well. They set up a presentation and they're fully prepared. They understand the client's business and financials. They've got the facts and figures. They rehearse the meeting and test their assumptions. They've dotted the Is and crossed the Ts.

Scenario 2

Don Levine gets a call from his manager, Wilma Wilson. Wilma needs Don to step into a deal with a customer he has not dealt with in years. Why the sudden call? The current sales rep has been reassigned to Alaska. Don is a pro, one of the top sales executives in the company, but he needs three weeks to prepare for this specific assignment. After all, this is a very important negotiation and it has been some time since he's dealt with this account. Unfortunately, Wilma gives Don two days to get it together, because she has already set up the call with the customer.

Don springs into action. First, he calls the former sales rep, but the rep has already left for Alaska to manage a new account. Undeterred, Don quickly calls other colleagues who have dealt with the customer. However, Don still can't piece together everything he needs. His knowledge isn't complete. Don knows he's not prepared, but he's done the best he can do in two days.

At the meeting, Don feels a lack of confidence. He answers questions with hesitation, hemming and hawing. It comes through that he's not as prepared as he could—or should—be. For most people, a lack of confidence comes through to the other side. If Don doesn't seem confident, how is the customer going to feel confident? What do you think the customer concludes? "Let's show Don the door." He may lose the customer now and forever.

IF YOU'RE NOT PREPARED

We all want to be the key player in Scenario 1. However, the second scenario is far more common. We rarely have enough time to prepare, and we seldom get the luxury of sufficient time to do everything we would like.

The question is, what is the appropriate amount of preparation? To decide how much preparation you need to do (especially when time is short), do a quick cost-benefit analysis based on the size and importance of the particular deal. That analysis will be influenced by the amount of time available to you.

So, what do *you* do if you're not as prepared as you should be, in a situation like Don Levine is facing? Write your steps in the workbook.

STOP if you are using the Companion Workbook.

Exercise 6-2: Default plan B

Here's what we suggest:

> **Step 1: Call in advance to reestablish rapport with the former customer and to set the right expectations.**

Don Levine called the former sales rep (unsuccessfully) and anyone else who had been involved with the account. He did not call the customer.

Don should probably have called the customer to explain that he was on the account and to get acquainted. He could have told the customer he looked forward to working with them again. He could have attempted to get a sense of how the customer viewed this change. Would the customer object about the reassignment and the fact that Don has not been the salesperson on the account in years? Or the fact that Don is now the third salesperson on the account in the past year? Possibly. But at least Don would have honestly set the customer expectation for the meeting. And then he could begin gathering information about the customer's business expectations.

Step 2: Set the right expectations given your motivations and the customer's.

Set expectations and understand motivations and objectives. Whose motivations and objectives do you want to know? You want to know your own as well as the customer's. Even in a short time, you should be able to gain an understanding of what your customer and you hope to achieve in this deal. Consider what brought you to the table and what you want to get from the negotiation. This leads to the **requirements,** reflected in the terms that will be negotiated into the contract.

If you understand motivations and objectives—your own and theirs— more often than not, you can anticipate what the other side is going to do. That's preparation! When you have knowledge, most issues you discuss have predictable outcomes. Effective negotiators maximize predictability and minimize surprises. Remember, predictability allows you to prepare your responses, which allows you to look knowledgeable and be convincing during an iterative negotiation process.

Step 3: Be straightforward about what you do and don't know.

Don't fudge, cover up, or muddy the waters. If you don't know something, gather information from the customer and say, "I'll get back to you on that." Think back to Mark Twain's advice: "Always do right. This will gratify some people and astonish the rest." We know many people who speak just to be heard, because they believe that they will be ignored if they don't say something. But once you say something not based on knowledge, it's impossible to pull it back! And if you truly have something to say, you should have the comfort to know that you will get the opportunity to say it later. As the adage goes: better to be silent and thought a fool than to speak and remove all doubt.

Step 4: Commit to getting the answers as soon as possible.

Don't promise what you can't deliver. First you must find out the customer's true concerns. Make sure that you understand not only what they are asking, but why. Customers seldom expect you to have all the answers at your fingertips, but they *do* expect you to produce the answers within a reasonable period of time.

Step 5: Use your time wisely.

Don't panic. We have seen far too much panic in the process of sales and negotiation. While panic may be a natural reaction to pressure situations, it wastes time and energy, and often results in incorrect answers. Instead, listen to the issues raised and prioritize them. Then set an agenda to resolve these issues. And if you feel pressured at a particular moment in a meeting, take a short break for an internal discussion. Keep in mind that your counterpart is usually as short of time as you are and they won't have all the answers either.

TIME FLIES WHEN YOU'RE IN A BIND

Imagine that you have only two days to prepare for a crucial negotiation. You need two weeks, but the time is just not there. Complete the workbook activity to order your priorities. Start by looking back at the five steps you already listed in the last exercise. Consider the steps we outlined. Then, place your priorities in order from most to least important.

STOP if you are using the Companion Workbook.

Exercise 6-3: Rank and file

Hard work gains you knowledge. In order to get your job, you first had to get an education. You might have a degree in engineering, math, science, marketing, finance, or accounting. The knowledge that you acquired from your education and related experience got you into that comfortable chair. You would never have been hired if you didn't have the "right stuff".

Similarly, you must prepare for each individual negotiation. Never underestimate the other side. Always assume they have done their homework and that they are prepared.

PRINCIPLE #2: PROTECT YOUR WEAKNESSES; UTILIZE THEIRS

What are some potential weaknesses you and your team can face in a negotiation? List six of these weaknesses in the workbook.

STOP if you are using the Companion Workbook.

Exercise 6-4: Weaknesses

How closely does your list match ours? We see many teams are confronted with the following weaknesses:

1. Tight deadlines

2. Lack of patience

3. Insufficient alternatives (if you are in sales: not enough prospects)

4. Lack of understanding of the customer

5. Poor cash flow

6. Product or service credibility issues

Bonus weakness: Poor teamwork—susceptible to divide-and-conquer tactics by the other side.

Play to your **strengths**. Three of the greatest strengths to bring to a negotiation are:

- You and your skills

- Your team

- Your preparation

The toughest people to negotiate with are those who don't give you input or don't know their criteria for decisions. As a result, you don't know what needs you must meet. What problem are you trying to solve for the customer? How can you articulate value and get your customer (the decision maker) to make a decision in your favor? And when you can't talk about your value—your strengths—you are going to be in a battle over your weaknesses. This makes it easy for the other side to exploit your weaknesses. How can you overcome this kind of situation? Here are two ways:

1. Find the niche. Research the other side's business weaknesses. Take the time to speak with other members of their team. Find someone who can tell you what their needs are. You should always sell in the context of the customer's weakness. That will tell you what they value.

2. If all else fails, make an educated guess about their criteria. Put your

proposal in front of the customer. Let them know what your assumptions were and why you made them. Say, "This is our best estimate of your [take your choice] problems/ROI/IRR/ROA/revenue goals/profit goals/other/all of the above. Here are the assumptions on which we based this. Please tell us what else we should consider. We are doing our best to help you." More deals are lost by sales teams unwilling to take a risk on framing a customer's business case than by those who do.

There are also other strengths that you may not have considered. For example, be aware of your personal style. If you are a "morning person," try to set up important meetings before lunch. If you are an "evening person," set up those crucial presentations late in the day. Try to make sure you don't schedule an important meeting or sales call in the evening if you are a morning person, or vice versa.

Consider this scenario:

> *A few months ago, we were advising a client we'll call "Charlie" in negotiating an executive compensation package with the CEO of a privately-held company. Charlie was being hired as the company's chief technical officer. Charlie understood cash, but he wanted to participate in the company's growth through stock options.*
>
> *The company was offering him 500,000 options vesting over three years at the exercise price of $2.50 per share. Charlie thought this sounded like a good package, but he was nervous about accepting a deal that he did not fully understand. Being smart enough to know what he did not know, Charlie came to talk to us. We started with the following three questions:*
>
> ***1.*** *How many shares outstanding plus the option pool does the company have?*
>
> ***2.*** *What is the industry standard for equity participation for someone in your position?*

3. *What is the current market value of the company?*

Charlie hadn't thought of these questions, and he had no idea of the answers. So, he found out from the CEO how many shares of stock the company had outstanding and the size of the option pool. We did our own research and found out the industry standard for equity participation. We also explored different ways of valuing the company that would ultimately result in the value of the stock if and when he exercised his options. As a result, Charlie was able to go back to the CEO with data in hand and confidently ask for 750,000 options exercisable at $1.50 per share. He got it!

Charlie's strength? He knew enough to get the help he needed to cover his weaknesses and gain knowledge to succeed in his negotiation!

PRINCIPLE #3: A TEAM DIVIDED IS A COSTLY TEAM

As any child has realized, if you can divide your parents, you can conquer them. Teenagers, of course, have brought this talent to an art form. Unfortunately, so have some savvy customers.

If you allow customers to pick you apart, they will. In fact, dividing your team may even be their conscious strategy. As a negotiator, you *must* have a united front so the other side doesn't divide and conquer your team.

Consider this scenario:

One member of your team is exceptional with numbers. You have complete faith in this team member because of her education, experience and integrity. But one day, she makes a huge error. The lead negotiator from the other side catches it and smugly says, "Your bright number-cruncher is a liar. A LIAR! I want her removed from this deal." Now, you know your teammate made an honest mistake. She is not a liar, but her numbers were incorrect.

What do you say to your counterpart on the other side? Write some suggestions on the lines in your Companion Workbook.

STOP if you are using the Companion Workbook.

Exercise 6-5: Math wiz

Answer: Once again, "it depends" is an important phrase in negotiations. We take the following position based on the few facts provided: You support your team member, even if she made a mistake. Explain that to the other side, with an apology. People make mistakes in negotiations—we all have. That doesn't mean the other side won't exploit your mistakes to erode your **Negotiation Capital©**. It is in your best interests to work together with your team. "A team divided is a costly team," of course has its converse, which is, "a team united is a powerful team."

Stand by your team. The credibility you will gain for your internal negotiations and support is usually worth it! Ultimately, most customers will appreciate your principled stance. If teammates allow their team to be divided by the other side, how will you maintain your team's integrity? What about the support you need from your teammates? As it erodes, so will your position. Of course, this does not mean that you should keep your team members even when you lose trust in their ability to support a negotiation. But whether or not you ask someone to leave your team should be your (and management's) own decision.

Another manifestation of bad teamwork that we see frequently is teammates not operating as a unified team in front of the customer. When team members disagree and the disagreement is apparent to the other side, what do you think happens to the other side's perception of that team? If you can't operate as a team in negotiations, how will you deliver your services or products?

In either case, "Debate internally; unify externally!"

PRINCIPLE #4: CONCESSIONS EASILY GIVEN APPEAR OF LITTLE VALUE

Consider this scenario:

> *Customer says, "I want a 25% discount."*
>
> *You say, "Done."*
>
> *Customer then thinks, "Wait a minute. That was too easy. I should have asked for 30%...or 40%..."*

To paraphrase Thomas Paine: What we obtain too cheaply we value too lightly.

Concessions easily given appear to have little value. If you have to make a concession, you should make a **principled concession.**

A **principled concession** is a concession made with credible business rationale that the other side values. The best principled concession is a concession granted by reference to value. Because it is *principled*, it should not be perceived by the other side as a giveaway. A principled concession makes the customer feel they really got something. They must feel they have earned it and deserve it as a valued partner. Principled concessions help you earn trust and respect. As we have already seen, credibility helps you close issues faster and move closer to closing.

Some potential principled concessions include:

- A price reduction based on value assumptions that are not what you articulated

- Added functionality (or value) for the same or greater price to ensure the customer gets the value as articulated

- Extended payment terms to cover for value that will be delivered later than originally assumed

- Added support to ensure the customer implements the solution based on the assumed schedule

- A customer concession because of your proof that the value is greater than they originally assumed

BUT...it's not enough to simply give the principled concession to a customer. You should go one step further and prove the value of the concession to the customer step by step. The process looks like this:

A prerequisite to a principled concession is a credible offer. A credible offer is a firm offer based on rationale related to value. A credible offer does not change arbitrarily. It changes if the value assumptions change. Those changes are principled concessions. More on Principled Concessions later, especially in Chapter 11.

PRINCIPLE #5: NEGOTIATION IS A CONTINUOUS PROCESS

The following is an illustration about the battle over leverage based on value:

A salesperson wants to sell Harvey a spiffy new Mercedes with $2,300 seat warmers. Harvey has never owned a car with seat warmers. However, he has owned cars that have had electrical system problems.

So, Harvey says, "I come with my own heating system, so I don't need seat warmers. And, I have had electrical problems in the past and seat warmers are just another electrical problem waiting to happen. Please take them out."

Figure 4: The principled concession process

Even Harvey knows that it's unlikely the salesperson is going to rip out the wires to the seat warmers.

The result: Harvey got the seat warmers without paying an additional $2,300. He maintained his leverage by showing he didn't value the seat warmers. But Harvey grew to love the seat warmers, and has since bought several cars with this feature. (Of course, Harvey's affinity for seat warmers is a leverage point he manages.)

Harvey managed leverage by lowering the perception of value in this case. But did the salesperson increase leverage for future transactions by getting Harvey to find the new feature truly valuable? It's similar to what happens in the technology industry when sellers give away new products or features for their customers to try. Later, when the value of those items is acknowledged, they come with a price.

> *Successful business negotiations are a marathon, not a sprint. You're in it for the long haul.*

I recently spoke with a client who had just lost a major account that had been theirs for many years. When debriefing his counterpart about the lost account, our client's president found that in previous dealings with this customer, the company's sales rep was arrogant and inaccessible.

Now, when there was a competitive alternative perceived as "almost as good," the customer chose to go with the alternative.

Of course, there may have been other reasons for that decision, but clearly the sales rep's behavior when he had the leverage damaged the long-term relationship.

You're working on the ground floor of negotiating deals. Since your negotiation work now can result in a long-term future (or no future), you must regard negotiation as a continuous process. After all, establishing long-term business relationships makes doing business easier and more efficient for both sides. Even if the sales rep has short-term motivations because of his quota at that time, he may have ruined his personal reputation for future business with that customer.

The concept that a negotiation does not end with the contract signing is particularly true in the information technology industry. Many contracts are written before implementation takes place and not all requirements are known. Contracts themselves reflect that, and provide for continuing negotiations (usually referred to as "governance"). Most contracts have an amendment process, evolving statements of work, change control procedures, engineering change mechanisms, or provisions for out-of-scope requests.

Imagine that you and your customer have agreed to all the terms and the agreement is being signed tomorrow. Now answer these questions in the workbook.

1. Does the negotiation end when the contract is signed? If not, why not?

2. If not, when does it end? Explain.

3. Is there any term in a contract that does not have a financial impact? Explain your answer.

STOP if you are using the Companion Workbook.

Exercise 6-6: All's well that ends well

Regarding questions 1 and 2, we believe some of the most difficult negotiations occur after the initial contract is signed. Many deals in the technology business are made before the solutions are implemented or services are performed. By their very nature, those types of deals involve continuing discussions about products, evolving goals, solutions, services, delivery, implementation, the statement of work, etc. Discussions between customer and suppliers should never end, especially if the relationship is good. An end usually means the relationship is over and only the "survival" terms of the agreement are being implemented.

UNITED WE STAND

Why does getting together with your team and coming up with a thought-out negotiation strategy for a particular customer give you a better chance of success, than relying on your instincts?

1. The other side will have less opportunity to use "divide and conquer" on your team.

2. The best minds generate the best ideas, which lead to credible teamwork and to the best solutions.

3. Management is part of the team and will be satisfied with a supportive role, because they will be confident in your preparation.

Remember: Motivations, objectives and requirements will be different between customer and supplier as well as within the different functions of each party. Both parties know that. Therefore, the solution must satisfy the various interests of each party to a significant extent. And the value of that solution should be quantified to all concerned.

Each company will ask the question, "Are we an improved company by doing this deal?" and "Is this my best option for solving this problem?" If the answer is yes, the deal should close and the relationship should commence.

Pay special attention to your customer's consultants. What is their motivation? If the consultants are going to participate in implementing a solution, you may be able to influence them.

PRINCIPLE #6: TERMS COST MONEY; SOMEONE PAYS THE BILL

Every term in contracts and negotiations should be of some value. And it has some associated cost. As a negotiator, knowing the rationale for a term enables you to articulate its value and identify its cost. The value of the total deal is the aggregate impact of all the terms. If you don't understand the rationale, not only is your credibility impacted, but so is your leverage.

Suppose you are negotiating a private label distribution OEM (original equipment manufacturer) deal, representing the seller. Their standard agreement has a term in the contract that states: "In the event any part fails within the warranty period, the customer may return the part at the customer's expense and supplier will send a replacement part within three business days." When the customer sees this, she indignantly raises her voice: "I can't believe you would expect me to pay for shipping...my people test the parts and it's your part that fails! I won't pay shipping costs for defective parts!"

What if you say, "Wow, I don't know how that got in there? Doesn't make sense to me. OK, let's take that out."

How will the customer feel about that reaction? How credible is the agreement? Will they wonder how many other "throwaway" terms are in there?

Maybe they will need to re-read everything now. Maybe the seller is trying to slide things in. And what about your own credibility as a negotiator?

It is necessary to first explain the rationale for the term to assure the customer that the contract is rational and credible. So you as the seller might say, "Let me explain why this is in here. The good news is our parts fail infrequently, but if we were to issue credits for shipping, our entire receiving process would need to change, because only procurement can issue credits. And right now, no procurement personnel are assigned to the receiving dock for that purpose. I could find out what the cost of changing the process would be, but in the past, that wasn't cost effective, and that cost would affect your overall price. I am happy to look at it again, though, if you like."

The customer may not be persuaded, but at least you kept your credibility with this approach.

Remember, there's no "free lunch". If you cannot articulate the value of a concession, and, many times, if you can't quantify it, it usually comes down to price. And you may not get a buy decision at all! Price is tied to value in function and finance.

But value is always two-sided:

- One side means value to you. What are you giving up or gaining to make the deal?

- The other side of value is the value perceived by the other side.

If the other side perceives that you are giving up something of value, they are more likely to treat your concession offer as something of value to them. If you are not treating the conceded term as anything of value to you by at least explaining your rationale, the customer will not treat it as anything of value to them in the concession process. As a result, they will not give you something of value in return, since their perception is that you have not given anything of value to them. A principled concession based on your knowledge of the value of the conceded terms impacts leverage in your favor.

WHAT YOU LEARNED IN THIS CHAPTER

- K&R's Six Principles™:

  1. Get M.O.R.E.—Preparation is key to a winning negotiation.

  2. Protect your weaknesses; utilize theirs.

  3. A team divided is a costly team.

  4. Concessions easily given appear of little value.

  5. Negotiation is a continuous process.

  6. Terms cost money; someone pays the bill.

- Know what problem you are trying to solve—it's not always money.

- **Principled concessions** are concessions that have a credible business rationale the other side values. Because they are principled, they should not be perceived as arbitrary giveaways by the other side.

- A credible offer must be supported by credible rationale.

CHAPTER 7: K&R'S NEGOTIATION SUCCESS RANGE (NSR)™

"I can't get no satisfaction," the Rolling Stones moan, but you *can* often "get what you want" by negotiating smarter with the K&R Negotiation Method. To do so, you have to get into the zone—**K&R's Negotiation Success Range (NSR)™,** that is, the zone within which both parties will be satisfied with the deal.

FINDING THE NSR

Easier said than done. When you begin planning for your negotiation, the seller's prices and consequently the buyer's costs in the transaction are critical factors in almost all deals. In fact, most people perceive that price is a deal maker or deal breaker depending on their view of value. That view is dependent in part on how well you create leverage by articulating value, which we've already discussed and will continue to discuss further.

THE POTENTIAL DEAL BREAKER

The NSR helps you understand relationships between three key price points:

- Your initial offer

- Your walk-away position (lowest price if you are the seller)

- Your estimate of the other side's walk-away position (highest price you believe the buyer will pay, if you are the seller)

The initial offer is the first position taken with the other side. That can be in the form of you making the first offer or your counter-offer to their initial offer. These offers should be made with rationale that makes them credible. We will discuss credible offers more in this chapter.

Your walk-away position is your lowest ("best") price, if you are the seller. If you are the buyer, it's the highest price you are willing to pay, above which you walk away. Based on analysis of your infrastructure, costs, and your perception of the value of the deal to you, your team should agree on your walkaway position before the negotiation starts. That does not mean that position cannot evolve in the iterative negotiation process. In fact, it often does, particularly if either side is strong in making their value arguments.

ESTIMATING THEIR POSITION

Your estimate of their walk-away position (the customer's highest purchase point if you are the seller) is based on your information gathering about the other side. That means knowing about many important items such as: (a) the customer's business impact from your solution, (b) the customer's cost structure, (c) how they perceive alternatives to accomplishing their goals of this deal, (d) what their opportunity costs are, (e) how they are positioned in their industry, (f) how their competitors accomplish what your product or service delivers, and (g) other information we will discuss later. Your value proposition is built on this knowledge. So, in estimating their walk-away position, you must have a sense of how their business case supports your deal. Your goal in making your estimate is to get as close to the price they can afford if they "buy" your value proposition.

What is the relationship between your estimate of their walk-away and your initial offer? In our view, you usually want to make your initial offer somewhat higher than your estimate of their walk-away. Of course, any such offer needs to be supported by good rationale. Making the initial offer higher than your estimate of their highest price allows you to test your estimates without leaving money on the table. If you make your initial offer too far outside your estimate of their walk-away, the danger is that they will walk away without discussion. These concepts should become clearer in this chapter.

As a buyer, you would make the same estimates, but from the other side of the deal. In fact, you often have data from past deals and industry sources that can assist in estimating the seller's costs and competitive position against the values to you. There are a couple of key differences. First, your initial offer will be somewhat lower than your estimate of the seller's lowest sales price.

Second, you have a realistic option of making a "low ball" offer in order to get the discussion focused (or "anchored") on your end of the NSR. Why can you usually make such an offer without fear that the seller will walk away? Because in a normal supply/demand environment, sellers are not likely to walk away from potential business. That may not be the case in an environment where there is limited supply or scarce resources. Either of these limitations give a supplier leverage.

Whether you are a buyer or a seller, by understanding your three price (or terms) points (your initial offer, your walk-away, and your estimate of their walk-away) on the NSR and the reasoning behind them, you can begin formulating your negotiation and concession strategy to move the other side closer to their walk-away position. That moves them closer to your side of the NSR.

HOME ON THE RANGE

For a negotiation to be effective, both parties must feel they have succeeded.

For the deal to work over a period of time, the negotiation must have a Negotiation Success Range™. Here's a very simple example from the seller's viewpoint.

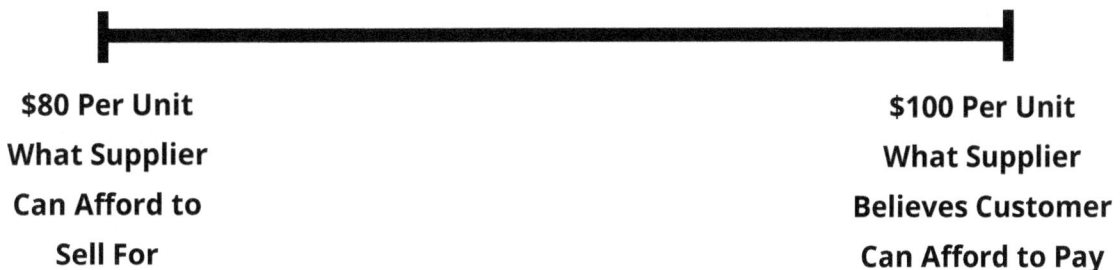

$80 Per Unit	$100 Per Unit
What Supplier	What Supplier
Can Afford to	Believes Customer
Sell For	Can Afford to Pay

Figure 1: Negotiation Success Range (NSR)

THE BUYER

If the buyer is purchasing at a price higher than or outside the Negotiation Success Range™, the buyer's cost will be too high. That's true whether the buyer is reselling or using the products/services in its operations. The buyer's competitiveness is negatively affected and he/she eventually stops buying.

THE SELLER

Likewise, if a seller is selling at a loss, the seller will eventually have to cut back by reducing quality, service or support. Or the seller will have to raise prices (if competitively able) to make up the losses. Otherwise, the seller has two basic choices: Stay with the deal and lose money or try to get out of the deal.

What's your job as a skilled negotiator? The goal of an effective negotiator is to get the other side to move closer to your side of the NSR. In other words, you want to induce principled concessions from the other side. Your principled concessions will make the other side feel good. You accomplish this by making credible arguments with persuasive rationale based on value.

K&R's Negotiation Success Range helps us visualize this process. We do it from the seller's viewpoint, but it works the same way from the customer's side.

As this chart shows, in a normal vendor/customer scenario your first offer as a vendor will be on the high end of the scale. The customer's first offer will be low. Notice that the two offers are often far apart. How can you bridge this gap? Read on.

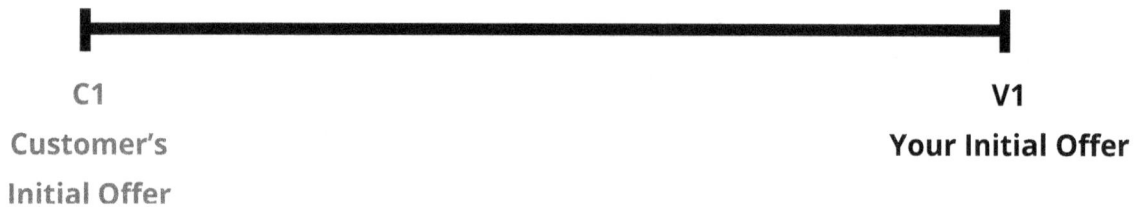

C1
Customer's
Initial Offer

V1
Your Initial Offer

Figure 2: NSR – Chart #1

Your Lowest Sales Point V3

Your Estimate of Customer's Highest Purchase Point C2

C1

Customer's Initial Offer

"Real" NSR

V1

Your Initial Offer

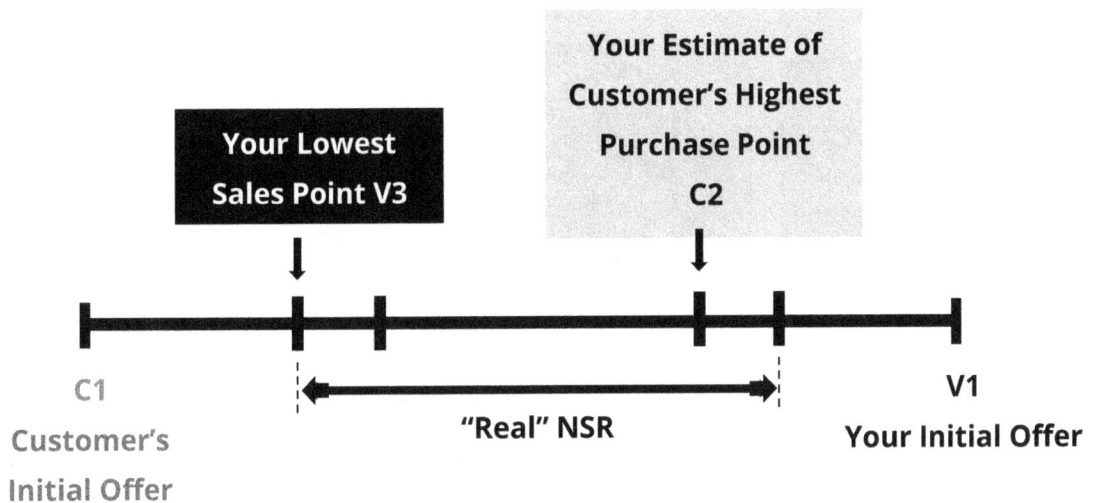

Figure 3: NSR – Chart #2

Look at Chart #2. Note how close your lowest sales point is to the customer's initial offer. As we mentioned, if the buyer has done a good job of estimating, their offer should be lower than your lowest (walk-away) sales point. The usual reason for this is that their goal is to have the entire negotiation take place on their side of the NSR. Marketing people often refer to this concept as "anchoring". Also note that there is often a vast difference between your lowest sales point and your estimate of the customer's highest purchase point. You still don't have the "real" Negotiation Success Range.

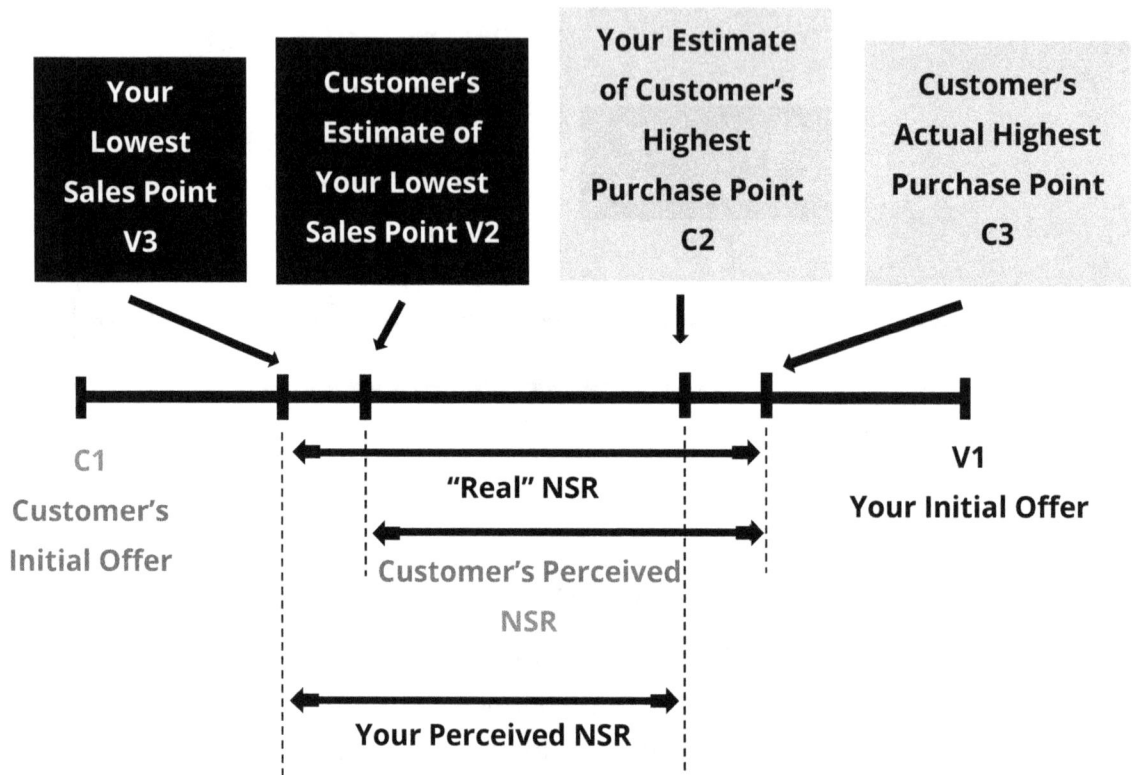

Figure 4: NSR – Chart #3

By combining the customer's perceived NSR, which is the customer's estimate of your lowest sales point and the customer's highest purchase point, you get the "real" Negotiation Success Range. This range lies between your lowest sales point and the customer's highest purchase point. Because one side of the "real" NSR is usually confidential to the other side, we never really know the "real" NSR. What we do know and can deal with is the perceived NSR, which, if you are the customer, is between the customer's estimate of seller's lowest sales point and the customer's actual highest purchase point, and, if you are the seller, between seller's estimate of the customer's highest purchase point and seller's actual lowest sales point. But it is up to you to keep probing. As a seller, you would like to drive the customer's estimate up as high as possible based on value. That means you must give the customer compelling reasons to move up to where you are—where you want them to be. It also means getting sufficient information so you are satisfied that your estimate of what they are willing to pay aligns with their walk-away. You use persuasive techniques to move the other side's perception of the NSR closer to your side. If this movement happens in a principled way and ends within the actual NSR, both sides are likely to perceive that they have won. Looking at this visually, the NSR prompts you to articulate the value arguments that will get the other side to move. After all, your estimate of their highest purchase point needs to be based on their business case, which is based on your value. Otherwise you have made a mistake in your estimate.

Consider the following scenario:

> *A vendor cannot sell below $20 per unit. Below that point, the vendor loses money.*

> *A buyer cannot afford to pay more than $19 per unit. Above that point, the buyer loses money.*

What might happen if the seller decides that he/she must have the business and decides to sell outside the Negotiation Success Range™? The seller sets the price at $18.50 per unit. Write your answer in your workbook.

STOP if you are using the Companion Workbook.

Exercise 7-1: How low can you go?

What we think is summarized a few pages back in "Home on the Range". Cutting service, support or materials are not the only options. Sellers occasionally sell at a loss as part of a "loss leader" strategy. When that happens, the seller, if contractually able to do so, may increase price after the buyer has become dependent on the seller's solution (such "low-ball" bids are discussed at the end of this chapter).

K&R'S NEGOTIATION SUCCESS RANGE (NSR) CASE STUDY

In the following case study, you can see the planning stage and actual events as they unfold. You can also see the cross-dependencies of terms upon price, and vice versa. As you read, track how the numbers fall on the Negotiation Success Range™. See how this spread influenced our strategy.

Here's the background:

The customer, a company buying our client's products for resale, had been provided with free basic support in the past. Naturally, they expected this benefit to continue. Our client faced growing customer support resource requirements, including new premium support services (PSS). We articulated the value of charging the customer for PSS support that would enable the customer to provide premium service to their own end-users for additional fees.

PLANNING STAGE

The following three charts show how we planned our strategy.

Our Lowest Sales Point V3

C1

Customer's Initial Offer:

- Free Support Resource

V1

Our Initial Offer:

- Support for Software and Hardware
- Annual Fee of $200,000

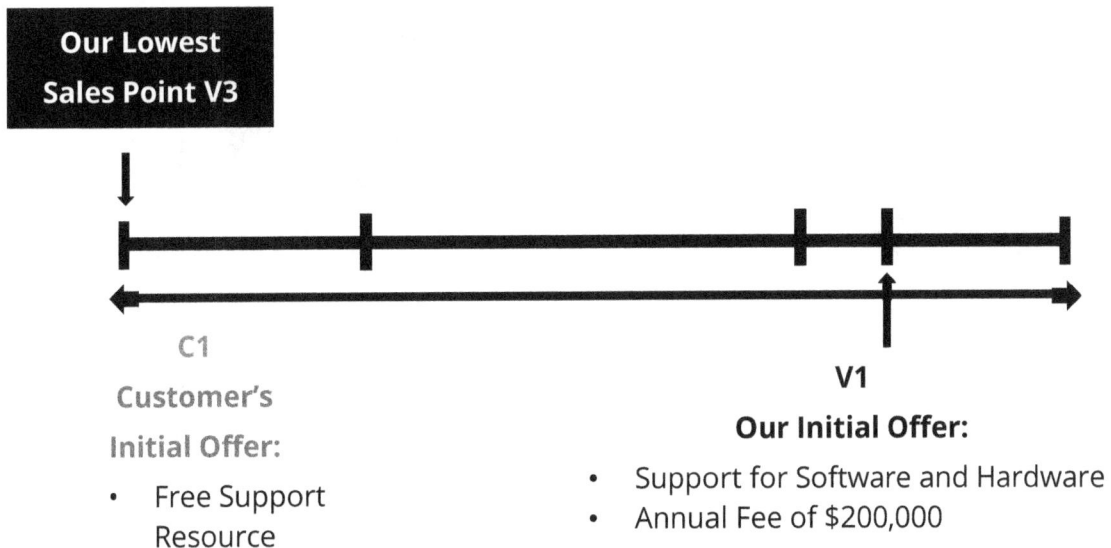

Figure 5: NSR – Planning Stage 1

During the planning stage, we considered the fact that the customer was used to getting the support for free as their initial offer, and as our lowest sales point. Our management would not walk away from the deal, so if the customer refused to move, the support would be given for free again. Our initial offer included the following: We would provide support resource for hardware and software, and charge an annual fee of $200,000.

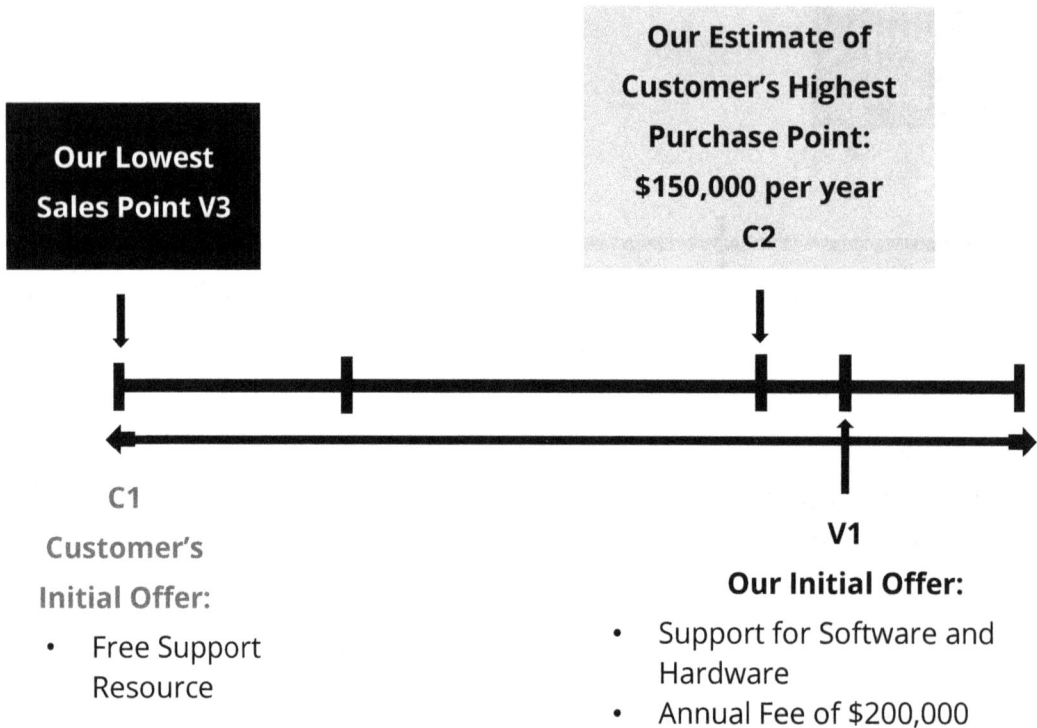

Our Estimate of Customer's Highest Purchase Point: $150,000 per year C2

Our Lowest Sales Point V3

C1

Customer's Initial Offer:

- Free Support Resource

V1

Our Initial Offer:

- Support for Software and Hardware
- Annual Fee of $200,000

Figure 6: NSR – Planning Stage 2

Then we added our estimate of the customer's highest purchase point. We estimated the customer's highest purchase point at $150,000 per year. Of course, all these calculations are based on articulating value, both qualitative and quantitative.

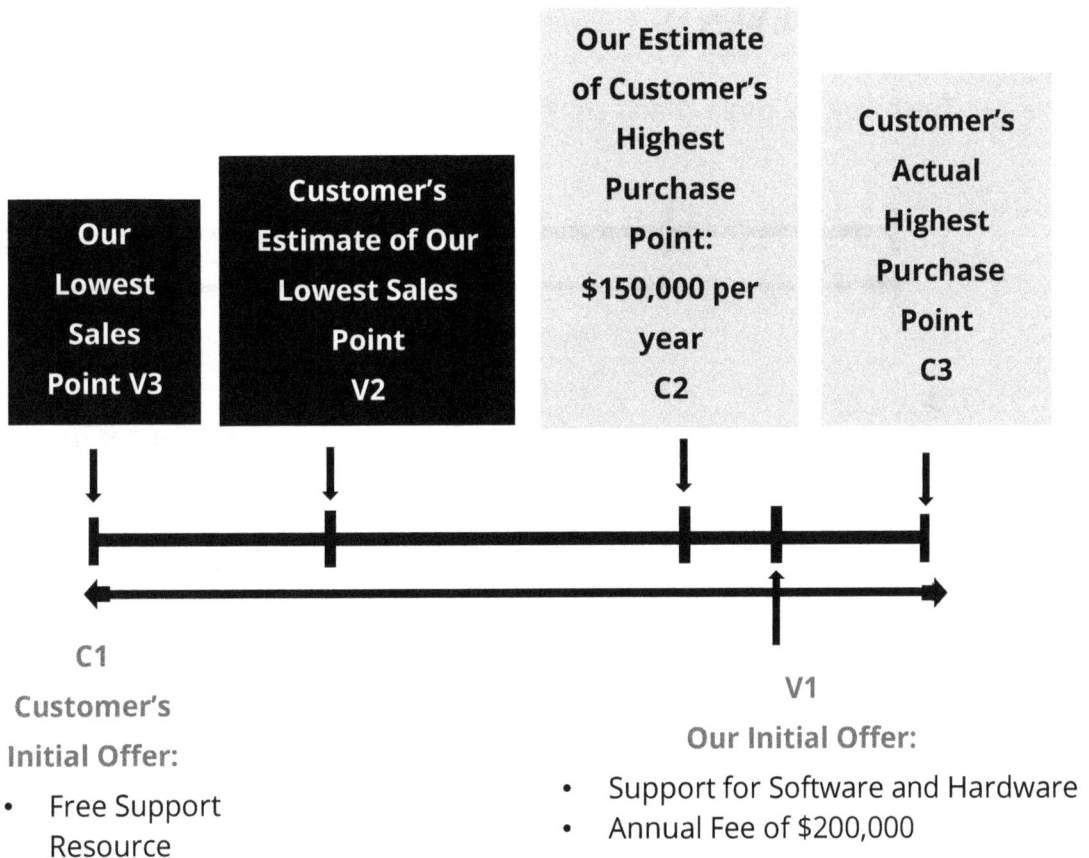

Figure 7: NSR – Planning Stage 3

Finally, although we did not know the customer's estimate of our walk-away point based on value we articulated, for the new premium support, we believed their estimate was higher than "free". For illustration purposes, we added in the customer's highest purchase point. Of course, we never really know the maximum any customer is actually willing to pay, so a confident estimate based on articulated value is critical.

ACTUAL EVENTS

The following three charts show how the negotiation actually played out:

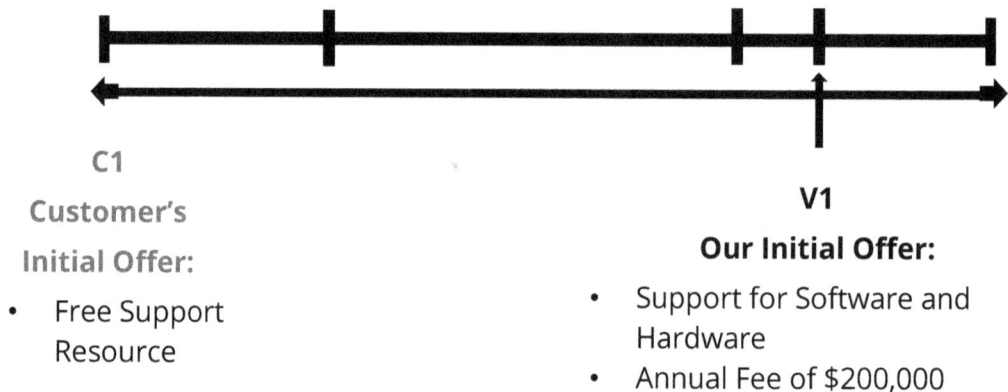

C1

Customer's

Initial Offer:

- Free Support Resource

V1

Our Initial Offer:

- Support for Software and Hardware
- Annual Fee of $200,000

Figure 8: NSR – Actual Events 1

We began with the same strategy we had planned out: We would provide support resource for hardware and software and charge an annual fee of $200,000. Naturally, the customer preferred to continue the past pricing: free support.

They made the value argument to us that providing premium support services helped their end-users, which would increase sales for all of us. As an alternative to us providing premium support, they asked to receive our source code for free so they could provide PSS themselves! They made the additional value argument to us that giving them source code would relieve us from having to provide premium support for their end-users. This was the term they were trying to use to influence price.

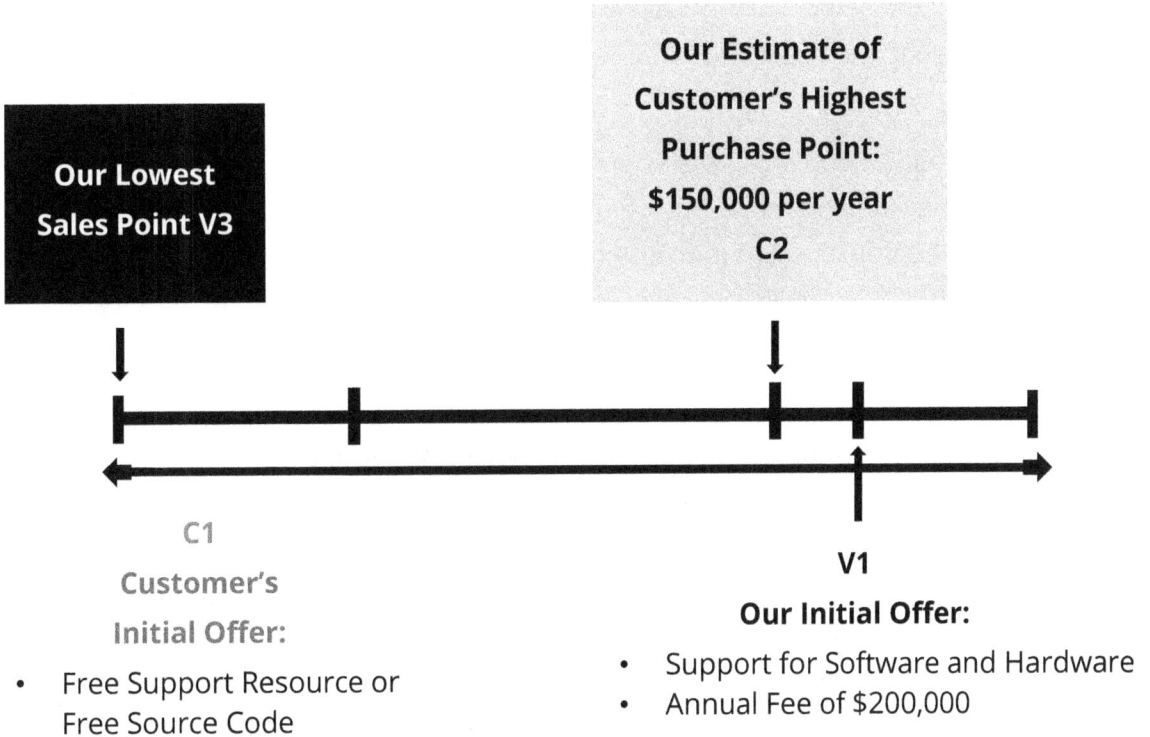

Our Lowest Sales Point V3

Our Estimate of Customer's Highest Purchase Point: $150,000 per year C2

C1

Customer's

Initial Offer:

- Free Support Resource or Free Source Code

V1

Our Initial Offer:

- Support for Software and Hardware
- Annual Fee of $200,000

Figure 9: NSR – Actual Events 2

We made the following arguments to avoid getting pulled down to their end of the NSR:

- Giving them source code was of no value to us. In fact, it would add risk to us by exposing a valuable asset.

- It's costly to support source code—$150,000 a year—since it would require additional technical information to be packaged for them and an onsite liaison.

- Training their people in support of our source code would require their resources and time at great cost to them. Because of our greater experience, they would still not match our support capabilities.

Because they asked for source code, we surmised that they were very risk-averse.

We felt we could use the source code request to pull them up to our end of the NSR and more. So, we offered them very limited source code rights as "insurance" to support their end-users. The price for this would be $150,000 annually. We repeated our argument regarding the quality and value of our premium support that they could pass through to their end-users for additional fees.

As a result of our understanding of where we were on the NSR, and the value arguments we made, the customer paid the annual fee of $200,000 and the additional annual $150,000 for limited rights to use our source code solely in the event that we failed to provide the required support. And, as a condition of providing the source code to them, they granted us unlimited royalty-free rights to any derivatives (improvements) they made in the process of providing their own support!

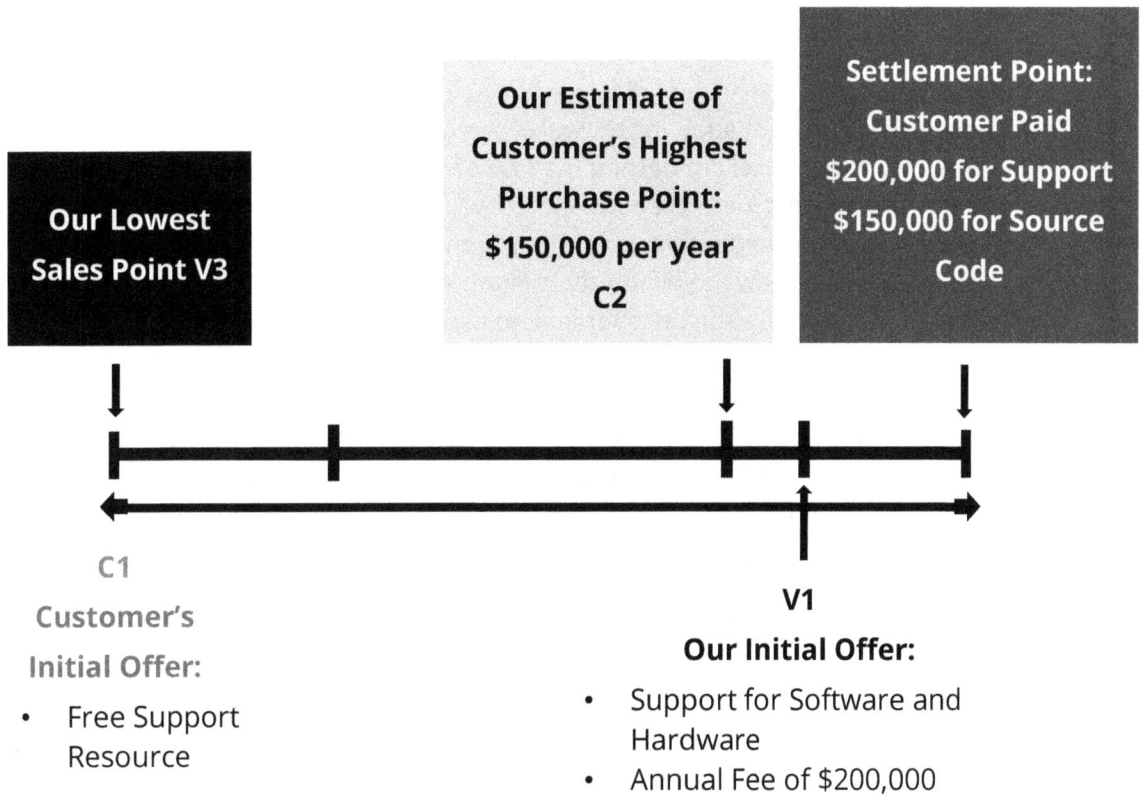

Figure 10: NSR – Actual Events 3

☞ The K&R Deal Forensic

This was a successful negotiation for several reasons:
1. Teamwork was excellent despite the fear of some on the team that charging a fee for resources that were previously "free" was risky.
2. Quantified value and rationale for the price were well articulated: Premium support services would grow the pie for the customers.
3. A good rationale for the price for limited use of source code was also given. We gave the customer what they needed as "insurance" for their end-users, for a price.

Bottom Line: Understanding the dynamic NSR™ forced us to think through and articulate value to pull the customer to our end of that NSR. In the end, a professional debate based on business rationales led to everyone's satisfaction.

CREDIBLE OFFERS

A credible offer is made with rationale and conviction. Judge the credibility of these approaches. Write your answers in the workbook.

1. Seller to customer: "I will sell you these components for somewhere between $10 and $13 apiece."

2. Seller to customer: "What do you think is a good price? I'm open for your input."

3. Seller to customer: "The software price is $200,000 plus $40,000 annually for maintenance. Please rest assured that our competitors are charging more for less functionality; here is the data you can check independently to verify what I am telling you."

STOP if you are using the Companion Workbook.

Exercise 7-2: The price is right

The first two offers are not credible because there's no business rationale or conviction. For the first offer, why is the range of price $10 to $13? Why not $9.99? $13.01? There is no rationale provided for the range so the numbers for the broad range seem arbitrary. The second case, where the customer is asked to make the offer, is more difficult. The seller may want the customer to make the first offer, but, as is often the case, it's not what you say, but how you say it. The statement of openness may be perceived as though the seller has no idea what their solution is worth! Only the third offer has clear logic and confidence. Because it has a business rationale, it is credible to those involved even if the other side is not persuaded. Remember, a credible offer by itself is believable. It is not persuasive until the rationale is tied to value to the other side. To make this credible offer a persuasive one, what does the seller need to do? The seller needs to state and preferably quantify the value (impact) to the customer who is buying the product or service.

WHO GOES FIRST?

You know that you must give the customer compelling reasons for them to move up to their highest purchase point. But who should make the first offer? Explore the risks and advantages of making the first offer using your workbook.

STOP if you are using the Companion Workbook.

Exercise 7-3: Tag, you're it

Chapter 7: K&R's Negotiation Success Range (NSR)™

Here's our advice: First, in most buy/sell situations the seller makes the initial offer. Why? Because as a seller, you often have a list price, an advertised price, or simply know your business better than the customer does. Making the first offer is okay as long as you understand the marketplace. Of course, when you do so the advantage is that you set the bar or anchor. But there is a risk in leading. If you are a buyer, your offer might be premature because you don't have the underpinnings or the knowledge that you need. As a result, your offer might be too high. If you are a seller and respond to a bid prematurely, you might bid too low and leave money on the table. Or, if you don't know enough about the customer and haven't first had a chance to quantify value, you take a risk that the customer will walk away if your bid is perceived as too high.

With a Request for Proposal (RFP), the buyer usually makes the first offer, setting the terms without setting the price.

WINNING IS UNEVEN

The concept of NSR is closely linked to the popular notion of "win-win". If the deal is within the NSR, it's by definition a win-win deal. But the participants rarely view the NSR through the exact same lens. For one, deals are not so evenly divided that each side benefits equally. And each company has a unique strategy to support their unique goals. For example, revenue and profit could be essential to one company, while market share is key to another. The measurement of success is often different, as is value. Understanding these different "lenses" can help you articulate additional wins for the other side, while maintaining your own.

BONUS: THE "LOW-BALL" BID AND REVERSE AUCTIONS

As a buyer, beware of "low-ball" bids designed merely to get the business. These low-ball bids often involve a plan to make up for today's low price by raising prices in the future. You can identify likely low-ball bids when you get a number of quotes and one is much lower than the others (for example, three bids between $50 and $55 and one bid at $26) for very similar products or services. The low price can be very seductive. You would need to understand why one vendor's price can be so substantially lower than others' before you make a decision.

If you are a seller confronted by low-ball bids from your competitors, it's up to you to provide the buyer with information about the risks associated with your competitors' offerings and the differentiators you provide.

We recently got involved with a reverse auction scenario on behalf of a client. The reverse auction is where multiple sellers bid a solution based on a buyer's specification, rather than the buyers bidding on a seller's goods or services. Our client, the seller, had lost round one on price, because they bid twice a competitor's price. Unfortunately, they had bid a full solution the customer would eventually need, while the competitor bid a basic solution that merely met the customer's initial specification. We advised the client to go back to the customer's procurement organization to make it clear that the low-ball bid by the competitor was for the initial spec only and would result in a much higher price for the full solution they would eventually need. To do that, our client had to provide the customer with documentation and value of the differences. As a result, the bid process was reopened (but not without a struggle).

WHAT YOU LEARNED IN THIS CHAPTER

- **K&R's Negotiation Success Range™ (NSR)** is the zone within which both parties will be satisfied with the deal.

- For a negotiation to be effective, both parties must feel they have succeeded.

- If both parties agree to prices and terms within the NSR, the deal has a good chance of resulting in a successful relationship.

- If the buyer is buying at a price higher than the NSR, the deal and the relationship will probably fail as the buyer's cost will be too high.

- If a seller is selling at a loss, the seller will eventually have to cut back by reducing quality, service, or support, or by raising price. The deal and the relationship will probably suffer as a result.

- The goal of an effective negotiator is to get the other side to move closer to the negotiator's side of the NSR by inducing principled concessions on the part of the other side. This is accomplished through persuasive rationale based on value.

- Making an initial offer can anchor discussions at your end of the NSR, if you know the facts, the industry, and the value to the other side.

CHAPTER 8: NEGOTIATION STEPS: INFORMATION GATHERING

Consider this history from our files:

We were representing a Japanese company, "Kao K.K.". We were working on a strategic alliance with "Green," who had a reputation for being difficult. Prior to meeting internally with the Kao executives, I called a friend who had dealt with Green in the past. It turned out that she had previously dealt with Green's lead negotiator. She advised us that he was very abrasive and confrontational and that he would maintain that behavior as long as he got what he wanted.

Understanding that our Japanese clients were culturally inclined to maintain harmony, we added some time to our team preparation before the first meeting with Green. We advised the Kao executives to let us handle any adversarial moments—that such moments did not require concessions, but required patience and logical reasoning, since this is the way the Green Company always behaved. Sure enough, in the first ten minutes of our meeting with Green, their lead negotiator said: "We are going to use our contract, and if you don't like it, you can leave!"

Our clients were shocked; however, based on our preparation, they let us handle the situation. I said to Green, "Why don't you give us some time so we can consider whether to get back on the plane?" We requested a separate room to have an internal discussion and consider the issue. Rest assured, we were not about to blow an important strategic alliance for our client over whose contract to use. But we wanted to send Green a message that would show them that their lead negotiator's arrogant and threatening behavior may have serious consequences, and that they needed to treat us with business respect. We also knew that the alliance was important to Green and that their lead negotiator would be in jeopardy if he blew this deal over such an issue.

Their lead negotiator proceeded to interrupt us a number of times for the next 45 minutes, asking whether we were ready to come back to the meeting. He was sweating a bit more each time. When we returned to the meeting, we suggested using a "joint" contract. This was a solution that was difficult to implement and which we would not recommend in the future. But their lead negotiator got the message and behaved more reasonably thereafter.

The K&R Deal Forensic

There is one very important reason why the Kao deal was ultimately successful: PREPARATION. Preparation resulted in:
1. Understanding the culture of our client.
2. Finding out about the other side's tendencies and behaviors.
3. Being able to use that knowledge to prevent making unprincipled concessions merely to respond to pressure tactics of the other side is essential to gaining business respect.

DIG WE MUST

Mladen says, "A lot of what we teach is based on psychology and much of it is intuitively obvious. Human reactions are predictable in most cases, but we are still dealing with probabilities, not certainties. You still have to make judgments in the field. But before you can make informed, conscious decisions, you must have the *facts*".

One of the keys to a successful negotiation is information. Let's explore how to gather information effectively.

Chapter 8: Negotiation Steps: Information Gathering

Note: For most experienced negotiators, this section should be a review. But as even professional athletes know, reviewing the fundamentals is often a necessary part of receiving good coaching. It's the same with negotiators.

Imagine that you are selling professional services, software and support to a managed health system. This is a very big deal, not only because of the initial order, but also because of the potential for a long-term relationship with additional sales in the future. Assemble your information using the chart in the workbook.

STOP if you are using the Companion Workbook.

Exercise 8-1: Getting the goods

Compare your answers with ours. Please note that our answers are not arranged in order of importance, and each of us will have answers not on the other's lists. The key is to be thoughtful, so we gather as much pertinent information as possible to enable us to make well-informed, value-based negotiating decisions. To some extent, of course, *whom* you want information about overlaps with the *kinds* of information you want.

PEOPLE TO NOTE

Whom do we want information about?

1. The company's organizational structure

   Knowing the chain of command is a good start to help you identify who probably has the actual power to close the deal. Identifying the person with their finger on the button also helps you forge alliances and not waste valuable time. Remember that you still need to identify the rising star. Oftentimes, individuals with lesser titles can be influential decision makers because they are connected to management, or they might be highly regarded and aggressive.

2. **Decision makers within the company**

   Whether or not you have access to the president's office, the decisions made there impact you. You may never actually deal with the people who make the ultimate decisions, but knowing who they are helps you tailor your value statement to their needs. What you say is likely to be carried to them—informally as well as formally—so you must understand the actions required to get leverage with the decision makers

   As you learned in Chapter 6, "Negotiation is a continuous process." In most cases, the deal isn't over when the contract is signed because you're not just making one deal; rather, you're setting up a relationship, a process that should be enjoyable for most professionals. One of Mladen's friends, a senior executive, is a teetotaler (does not drink alcohol). Whenever working on transactions, he always finds out what the other side's lead negotiator likes to drink. Mladen's friend then sends the lead negotiator a bottle of his or her favorite drink with the contract. It's a great way to cement a relationship, and more often than not, the contract comes back signed.

3. **The line of business**

   Knowing about the line of the customer's business that will be affected by your products or services helps you shape your value arguments. By knowing their business, you know with whom your argument will resonate and who is potentially an additional advocate of—or detractor from—your products, services, or solutions.

4. **Anyone else in the company who has influence over or is impacted by the potential outcome of the deal**

   You've got the obvious people, such as the ones you meet with. But people not even in the room can often exert a surprising amount of influence over a deal. For instance, a lower-level tech might say, "It's a great deal, all right, but we won't be technically ready to implement it for at least six months, so it doesn't do us any good at all." Knowing this helps you understand both barriers to success and their weaknesses. So you can create a solution utilizing their weaknesses to create your value. Remember, it is difficult to create an effective solution when you don't know what problem you are trying to solve. "Protect your weaknesses; utilize theirs." (Chapter 6)

5. Independent parties (consultants)

Companies paying big money for consultants are likely to listen to their advice. So, you want to convince the consultants of your value. Then, when they advise the company, they will have in mind what you bring to the table. Ask yourself:

- What are the consultants' motivations and objectives?
- How can I help them meet those motivations and objectives?
- What customer needs will the consultants focus on?
- How would meeting the consultants' motivations and objectives get them to value my ability to fulfill customer needs?
- How can I meet those needs?

You need this information, especially if consultants wield a lot of power within a company.

6. People in procurement

Since they issue the purchase orders, they have power. People in purchasing are always looking at the bottom line, so cost is paramount to them. But as you learned in Chapter 6, "Terms cost money; someone pays the bill."

Chapter 8: Negotiation Steps: Information Gathering

How can you make a value statement to the people in procurement? Having the facts about their motivations, objectives, and requirements helps you do this. Many procurement organizations use other measurements like satisfaction of their internal customers or risk management (particularly government agencies). You can often use these interests to make your value arguments and sustain price.

7. Other stakeholders

They have a stake in the deal, so they have a say in the deal. (Good examples are venture capitalists, investors, shareholders, banks, or financial institutions.) Therefore, you want information about them and about their needs.

8. Clients or end-users

Who will ultimately use your product or service? How will they use it? Where will they use it? What are they currently using for the same purpose? Knowing the answers to these questions helps you decide which features to include, which support is necessary, and which principled concessions will build you the most leverage. This information helps you ultimately articulate and quantify value.

When you think about whom you want information, also consider the *personal* motivators. A story from Mladen's files illustrates this:

> *I was about to leave the office of the CEO of a well-known software company (let's call it "Bigg"), when he asked me to stay and listen to a member of his team debrief a deal they were working on.*
>
> *When the team came into the meeting room, it looked as if the last meeting with the other side hadn't gone particularly well. The CEO had asked his team to negotiate a license for some software code from a software firm we'll call Small. Small was a growing three-year-old company with $10 million in annual revenues but running in the red. Bigg wanted to use the code for a single but important function that*

was missing from a turnkey solution that Bigg already had in the marketplace. Bigg didn't need Small to support the solution and was willing to pay $3 million in a one-time license fee. Because the use of the code was completely unrelated to anything Small itself was doing in the market, and because Small would have no continuing responsibility for maintenance, the $3 million was, in effect, pure profit for them.

Bigg's CEO thought that Small would be delighted with the deal. Growing companies that are losing money tend to be cash-strapped; and for Small, the $3 million should be just like receiving funding. However, as the negotiation team reported, Small said "No."

Bigg's CEO asked, "Did you explain that we are providing them pure profit? That this is better than third-party funding for them?"

"Yes, we did, and we told them they would not be responsible for maintenance. We told them our use was completely different than theirs, so this would be purely incremental revenue to them. They agreed with all of that...and they still said no!" was the reply.

I was listening and interrupted this exchange with a simple question, "With whom are you dealing?"

The team lead replied, "Small's president."

"What do you know about this president?" I asked.

At that point, Bigg's CEO was getting impatient, and said, "Just offer them $4 million because we have to get this deal done quickly." And so, they did.

After Bigg's negotiation team got rejected again, I was asked to help with the situation. In doing a little research related to the questions I originally asked, we discovered that Small's president had come from the venture capital company that originally funded Small. Previously, she had successfully built companies that she sold and, in one case, had taken public through a successful public stock offering (IPO). She was more interested in establishing her company as a credible growing concern that others would want to invest in over time, making Small's stock more valuable. Then, she and her venture capital partners could "float" the company on a public stock exchange or sell it at a premium to another company in the industry. To do that, she wanted a deal that would show growing revenues, over time, not a one-time blip of revenue. What this deal needed to succeed was a restructuring, not an additional million dollars.

Due to a lack of knowledge about who Bigg was dealing with, the CEO threw money at the problem when money was not the issue. Now that we knew who we were dealing with and what her objectives were, restructuring the deal became easier. But taking the $4 million off the table was difficult. By promising to make a "partnership" press release—adding further credibility to Small as a company with a future—we were able to negotiate for Small to agree to a lesser payment in a royalty-stream over time.

The K&R Deal Forensic

What did we learn? The team made one major blunder that had several effects.
1. The Bigg team did not know anything about the person they were dealing with.
2. So, they did not understand her motivations (having venture capital investments) and objectives (raising the stock value of the company).
3. As a result, they threw money at a problem when money was not the problem (requirements).

Part of preparation includes not only knowing about the company, its financials, its business, etc., but also understanding the motivations of the people on the other side of the table.

Always consider: Who will benefit personally from this deal?

FACTS ON FILE

What kinds of information do we want?

Of course, not all information has the same value, but every bit of information you gather is important in its own way. No matter how small the detail may be, file it away in your memory and your notes. It may be just the fact you need to make your value argument. Having information helps you reduce the other side's alternatives; not processing new information reduces your own.

We do to ourselves what a car dealer does to us. As a car buyer, we almost always have leverage (the money and alternatives), but a car dealer erodes that leverage by asking all kinds of questions that narrow our choices. These include: How much are you willing to spend? What color do you want? Who will be driving the car? What features do you need? Pretty soon the car dealer has you narrowed down to the only car of its type in the entire country, a car they happen to have!

In the same way, when we find a solution to our problem, we convince ourselves that we don't have alternatives; only that solution will work. Even though other solutions may work, we think they won't work as well. Since we found a solution, we often stop processing new information, handing the other side leverage.

As you prepare for a negotiation, you need to know...

1. The customer's weaknesses in their business

    Mladen says, "You sell in the context of customer weakness. Weakness comes in many forms: the industry, business, promises that public companies make to shareholders, etc.

    "Many weaknesses are simply statements of direction the company has made in its annual reports, promises that it has not yet delivered. The best sales presentations factor in how the customer's weaknesses will be lessened by the deal."

2. **How decisions are going to be made in their company**

If you don't know the decision-making process, you are exposed. First, you are likely to set unrealistic expectations with your management regarding closing the deal. Second, the deal will be eroded by someone on the other side who you didn't expect would have a say.

Knowing their decision-making process, as well as their goals, weaknesses, and strengths helps you form alternative solutions and a sound negotiation strategy. The more solutions you can create, the better your chances to protect your weaknesses and use theirs.

3. **Internal office politics**

"I don't play politics," you may scoff. "I'm a straight-shooter." Well, that's good, but some people *do*. Office politics are inevitable. Office politics are the strategies that intelligent people use in both public and private organizations to maintain a competitive advantage in their careers.

Office politics don't have to be smarmy and deceitful—they can be the subtle and informal methods that successful people use to gain leverage and power. In many cases, office politics can explain the corporate fate of equally talented people.

Office politics include:

- Understanding who has the power within a company—and who doesn't

- Getting access to people with power (based on *their* interests) to help you close deals

- Having the ability to control people or resources

- Getting other people to do things you want done

- Building mutual understanding and respect while negotiating

- Knowing ways to remain flexible and make concessions without losing face (principled concessions)

- Using techniques to handle unfair situations with confidence and diplomacy

- Mastering good social skills

Knowing the internal office politics of a company can give you a valuable edge in a negotiation because it can help you get information from the right people and to the right people.

4. External politics—the factors in the economy that impact buying

This means understanding the following:

- How economic, political, and other conditions impact your customer's industry in general

- How these conditions impact your customer specifically

- How these conditions impact your customer's perception of you and your industry

Let's consider how technology solutions from a large company may be regarded when the economy is flying and small companies are growing. The small competitors would say, "Our solution is leading edge; the large company cannot respond to your needs with the nimbleness that we can. Look at how slowly they move and change."

But when the economy is in recession and small companies are going bankrupt or can't raise capital, as was recently the case, larger competitors could say, "You can rely on the stability of our solutions. Look at those 'leading edge' companies of a few years ago. Most of them are not around now. We have supported our customers through many ups and downs; you can rely on us."

People's perceptions of risk shift dramatically between good and bad economic times. Current political circumstances also factor in.

You *must* know how politics and the economy influence you and your customer when you enter a negotiation. You also must know whether or not a company actually has the money and the overall ability to make the deal!

5. How people in the company are evaluated; What are their and the company's Key Performance Indicators (KPI)?

   Knowing how people are measured for bonuses, rankings, commissions or promotions helps you determine personal motivations that are often as important in getting a deal as the company motivations.

   Often the biggest personal motivation is simply making their job easier. Mladen tells this story:

   > It was the middle of the fourth quarter and the software sales team of Brush Co. was trying to sell a software monitoring solution to a major financial institution we'll call "FinCo". The customer IT executive had been convinced a few months earlier that the solution would help them tackle certain technical issues in servicing their international accounts. The Brush Co. sales team was getting pretty frustrated at a sale that should have been easy. They could not get FinCo's procurement team to move with urgency.

   > We at K&R were asked to help. I asked what the last involvement of the IT executive had been. The answer was that the sales team last saw him a month ago. He was very busy, even overwhelmed. I asked how he gets his approval for expenditures. "He has to go to the CFO for anything

between $500K and $5 million," was the reply. "Well," I asked, "Does he have a business case to show the CFO?" "Not as far as we know," I was told. The conversation continued like this:

"Can we build one for him?" I asked.

"We could, but we are not sure of all the FinCo data."

"Why don't we try with our best data and then give FinCo our assumptions?"

A few days later, with the assistance of an astute financial person at Brush Co., we built a sample business case for the IT executive, using standard industry KPIs. A meeting was scheduled with him for a week later. At the meeting, the Brush Co. sales rep showed him the business case built on credible industry data, which contained three alternative views: an aggressive, a middle, and a conservative effect of implementing Brush Co.'s technology. The conservative case showed a 240% ROI over two years! The IT executive debated and corrected some of our assumptions and asked for a copy of the business case. "This saves me a lot of work," he said.

Two weeks later, the IT executive had the CFO's approval and the deal got done in the week before New Year's.

The K&R Deal Forensic

There are several important reasons why this deal got done within our client's time frame:

1. Understanding and dealing with varying motivations is always important.
2. The IT executive didn't have the time or resources to do his own business case work. Brush Co.'s team made his job easier by doing the business case for him.
3. The CFO was interested in returns on investment, and that's what the business case showed!
4. Most importantly, we understood the industry KPIs, and were willing to use logical reasoning to build our business case, even when we didn't have all their data. Provided you make rational assumptions and articulate them to the other side, you should be willing to build a business case. If they start debating the assumptions, you are in a good place. Remember that your data about the other side will rarely, if ever, be as good as theirs.

Understanding individual and company motivations help you design a value-driven proposal that gets results.

6. The company's financial status and stability

   Whether you are a buyer or a seller, you want to know the financials of the company you are dealing with. If you are a seller, obviously, you like to get paid. No one wants to be holding an unpaid invoice when a supposedly healthy company goes belly-up.

   On the other hand, if you are a buyer, you want to know that the seller has the financial capability to support you over time, especially if you are going to be dependent on their long-term maintenance and support.

   We have occasionally heard procurement people say, "Don't worry about the seller's financials, since this is a commodity. Let's get it cheap while they're desperate." Think about this: You get the least expensive commodity, let's say computer memory, from a failing company. Because of their financial troubles, they skimp on materials. You put their memory in the machines you sell to your customers worldwide. Now your computers are failing, and the cause is faulty memory chips. Your explanation to management is: "We got the cheapest source of supply...it's easy to switch...it's a commodity." We don't think this will fly, do you?

Sometimes it can work to your advantage to enter into a deal with a company that's not financially stable, but you'd better go into that deal with your eyes wide open. For example, if you are a seller, a buyer in bankruptcy reorganization can only spend money they actually have. In those cases, you can still sell to their needs. You just need to do it on a cash basis or using letters of credit. Just check with your finance people and legal counsel first!

Does this really happen? You bet. One of K&R's principals did just such a deal for a client with a company in Chapter 11 bankruptcy reorganization. The deal was for $14 million worth of products over three years. It enabled the other side to come out of bankruptcy. Yet even after they exited bankruptcy, the same payment terms requiring cash or letter of credit remained in place. We knew that our client's finance people thought it was great!

7. The company's timetable for purchasing

To some extent, the customer's timetable is dependent on the perceived value of your solution. The greater the customer's need to have a particular problem solved and the more immediate the impact of your solution, the more accelerated the customer's timetable will be. (Remember "Protect your weaknesses; utilize theirs!")

You need to test the customer's perceived timetable early in the process.

Factors influencing it are: The extent of the problem being solved, other priorities that require budget dollars and the relative position of those priorities from a business unit, and the company perspective. If the priority of the problem is not high, part of your job is to increase the perception of its importance on the priority list.

We were helping sell a technology solution to a budget-constrained customer. While the customer was spending some money, the priority list was short. However, there were items such as new office furniture in the budget. By pointing out the productivity gains to the administrative staff from the technology solution (quantifying and articulating value), we managed to push the technology purchase to that year's budget while the furniture purchases were postponed. Our ability to do this began with knowledge of the customer's initial priorities and timetables.

In a busy world, knowing when a company is most likely to sign helps you set priorities and save time and effort. It also helps you calculate the Negotiation Success Range™ with reasonable accuracy. Even though you know the company won't be purchasing for months, business pressures often create a tendency to throw money and concessions at them to hurry the buying decision. Most often that money is thrown away, since money is not the problem (as you learned earlier in this chapter). The deal has a better chance of getting done sooner based on quantified value arguments.

8. The company's history

Companies, like people, have a past. Knowing the history of a company—especially its past dealings with your company and your friends and acquaintances at other companies—can help you avoid pitfalls and build on existing alliances.

And think about this: suppose on the second round of discussions with the other side, their lead negotiator says to you, "Look, why don't we just use the payment terms from the deal that I did with your company three years ago?" You don't want to be in the position of having no idea what that deal was. Would your credibility take a hit? More than likely. Of course, that depends on how you handle it, but it would have been great to know about the previous deal in advance. Knowing about the previous deal can indeed be a shortcut to getting this deal done, or it can provide you with rationales why you don't want to use the terms from that deal. Either way, you are better off than having to fumble for an answer. Being as completely prepared as reasonably possible given the importance of the deal and the time available can turn potential credibility hits into credibility gains.

9. Other useful facts about the company

We've covered the obvious stuff you must learn: power structure, chain of command, office politics, financial stability, approval process and more. But don't stop looking for information and listening: get as much information as you have time to gather. In a relationship business you should keep processing new information even when the initial negotiation is over. Remember: Your best business partners for the future are your current ones. Keep your eyes and ears open, because even if a fact doesn't seem relevant at the moment, you never know when a specific bit of information could prove useful.

10. The other side's general business and industry trends

Following industry trends is invaluable. After all, as we already mentioned, we sell in the context of the other side's weaknesses. To understand their weakness, we need to understand their relative position in their industry, their business initiatives, promises they've made to their customers and shareholders, and other similar information.

11. Your *own* company's decision-making policies

Don't underestimate the importance of knowing your company's internal decision-making process and the motivation of the players. We will talk more about this in Chapter 15, but take a moment now to think about what happens to your credibility and leverage when you get blind-sided by your own team, all because you didn't know your own company's process or the participants' motivations. Before the process starts, reach agreement with your team and your management regarding how you will work together.

HOW MUCH INFORMATION IS ENOUGH?

Read the following two scenarios:

Scenario 1

You've been asked by your management to step into the Dataco negotiations because Curley, the previous lead negotiator, has left the company. A meeting with Dataco has been set for two days from now. Your management specifically asked for you because you have negotiated with Dataco in the past. However, you know nothing about this deal, nothing about the specific customer needs or the solutions involved, nothing about the issues, and nothing about the motivations of either side. Your management is unwilling to postpone the meeting because timing is critical.

You do the best you can in the two days you have, but you still have many unanswered questions and nagging doubts about your role as you walk into the meeting with Dataco. You're not feeling confident. Does your lack of confidence come through to the other side? Maybe you are more hesitant than normal in responding to questions. You are unsure of your rationales. The result? If you lack confidence in what you are saying, how can you expect the other side to have confidence in what they are hearing? Your credibility may be severely impacted!

Scenario 2

> *You've been asked to prepare to negotiate with Dataco, to enter into a licensing relationship for their technology. You know their company from prior negotiations and have a good grasp of the technology involved. You also have plenty of time to prepare for this negotiation. You use your time wisely to analyze the business case for both your side and theirs.*
>
> *When you enter the conference room to begin serious discussions with Dataco, you feel confident. Even if you can't control the competitive atmosphere, you feel confident about what you can control—your knowledge base. As you respond to inquiries from Dataco, you exude confidence. You answer questions and present your arguments with knowledge and rationale. The result? When you seem confident and that confidence is rooted in knowledge, you appear credible, and the other side is much more likely to believe your arguments.*

We often ask: "Is there any information you don't want?" People have trouble with that question. Maybe because the answer could be phrased as: "How do we know we don't want it until we hear it, and then it's too late...we already have it!" Of course, we make conscious decisions not to take in certain proprietary information that comes with responsibilities we don't want to have. Otherwise, all legally gained information is fair game.

Let's face it: You're never going to have all the information you want. There's never enough time; you always have too much to do. But relax. You'll have the amount you need to feel comfortable relative to the importance of the deal in front of you. So, stick with us; there are more great techniques for information gathering coming up!

Chapter 8: Negotiation Steps: Information Gathering

GETTING TO THE SOURCE

What are some sources of information?

We've shown that you need the facts. How do you get them? There's nothing mysterious about the process. Try some of these ideas.

1. Study past deals and contracts

You can start by checking your own company's files. Learning the history of previous deals helps you anticipate what the customer will be asking for this time.

As you study past agreements, realize that the other side will recall the terms and facts that favored them but conveniently forget the terms and facts that favored you.

2. Consult with fellow employees, past and present

Talk to the other side's team, current employees, and even former employees. Retired employees might be willing to share information. This population can be an especially valuable source of past deal history, and they can often be reached through mechanisms like LinkedIn.

3. Attend trade shows

Trade shows are a strong resource. Go listen to your competition's presentations and attend their break-out sessions. Likewise, attend your customer's trade shows. You can find out a lot about their strengths and weaknesses in the industry. That will help you shape your value arguments to them.

4. Read the company's annual reports

Annual reports from publicly traded companies reveal an astonishing wealth of information. Start with the financial reports. Look at the charts and numbers to find trends: Are sales up, down, or steady? See especially what the company views as its strengths. You are going to have to work very hard to sell to an area that they see as their strength.

Also look at their weak points, usually called "challenges". These are very important areas where you can often build a strong value argument. In fact, the promises of company initiatives tend to deal with the company's weaknesses. The statement of initiatives in an annual report contains some important weaknesses, because those initiatives have yet to be fulfilled. Think how your business case will look, especially to a "C" level executive (CFO, CIO, CEO), if you add a few references to items that the company promised their shareholders. And how will that affect your credibility?

Read the actual words, and then "read between the lines". Make inferences (educated guesses) to see what the company is *not* saying— and why. Look to see which division has the splashiest pictures. Who has the most text space? Which divisions are buried in the middle of the report? From what is actually in the report and from what is *not* in the report, you can infer which divisions are the healthiest. This often leads to larger budgets, where it's easier to make investments for improvement.

Much of the information about a company is available online in the Securities and Exchange Commission's database called EDGAR (for US traded companies). You might even want to buy one share of stock in the company. Then you'll automatically get all stockholder communications, including annual and quarterly reports.

5. Access a company's websites and the sites of their competitors

Since companies brag about their strengths, you can often use information on websites to figure out a competitor's weaknesses. If you go to the web page of your customer's competitor, you can see the differences between what the companies cover in their promotional material, and can deduce their respective strengths and weaknesses.

6. Read general business reports

How the company is perceived by the public tells you a lot about their strengths and weaknesses. That kind of information often comes from business news sources such as the *Wall Street Journal*, Dun and Bradstreet Reports, Hoovers, Harte-Hanks Reports, etc. Check them out on a regular basis to track companies you do business with now as well as companies you want to sell to or buy from in the future.

7. Look at pictures and memorabilia in the client's office

We all know the old saying: "One picture is worth a thousand words." What someone chooses to display in his or her office—which is essentially a public space—often reveals a great deal about the person. For example, if the boss hangs up a photo of a golf outing, you can likely conclude that spending time on the links—with you—could help advance the business relationship.

8. Access reports of industry experts

Independent industry experts often have key research about companies and industries. (They often know more about the customers than the customers know about themselves!) Get research reports on industries from organizations that you and your customers have relationships with. Major accounting firms such as Price Waterhouse and Coopers and Lybrand and Deloitte Touche Tumatsu Ltd., major venture capital firms, and consulting organizations such as Gartner can provide very good data in specialized industry sectors.

9. Wine and dine the other side

We mentioned earlier that in an informal environment where everyone is more at ease, valuable information can be revealed. Further, the bonds of relationships can strengthen. Keep in mind that in this environment, it is better to receive than to give (information)! Take your customers out to a meal and let them talk. Listen closely. Then focus on the subtext—what they're *not* revealing. Why do they really want to do this deal with you? What information has been revealed?

10. Talk to the "regular" workers

One superb sales professional we know brought donuts and coffee to the people working the press late at night. Softened up with sugar, caffeine, and exhaustion, the pressmen revealed important information about the company's stability and politics. Those inside the company can often reveal information in a minute that would take you days or even months to get from other sources.

11. Make a personal visit

Scope out the company yourself. For example, a crowded parking lot suggests a prosperous company, while a cavernous plant with few employees suggests the company may have serious cash flow or other business problems.

Ask the other side for a tour of their facility or business. Usually a firsthand visit is worth a thousand secondhand accounts! Here's one account from Mladen's experience:

> *One time I was negotiating with a software company on behalf of a larger client. The deal was very important for our client because the software company had some unique proprietary technology. For the software company, it was also important: The deal would equal 30% of their annual revenues. I asked for a tour of their facility. On the tour, our guide turned left, but I saw a corridor to my right and made a quick turn that way. Through the corridor, it was apparent new space was being constructed—essentially doubling the size of the software*

company's facility. I said to the lead negotiator, "What's going on?" He wouldn't give me a definite answer. He didn't want to admit that the company was gearing up to support the deal with our client. As a result of this discovery during the company tour, we gained confidence and leverage!

A picture is worth a thousand words. Visit and tour the other side's site if possible. You will gain a far better understanding of their needs and your true leverage position.

If you decide to make a personal visit and ask for a tour, you may want to do so on the spot to get a more realistic view of their operations. If the customer knows in advance that you're going to want a tour, they'll likely arrange for you to see what *they* want you to see—not necessarily what *you* want to see!

12. Become *their* customer

One way to scope out a company's strengths and weaknesses is to become a customer for their products or services. If the company is a retail outfit or a bank, go shop in their stores or open an account at one of their branches. It will be an eye opener. Whether or not you actually become a customer, they may be willing to let you attend the education sessions they provide for their customers (just as you provide education for your customers). What they stress is a key to what is important to their customers. And you may get a chance to interact with their customers to find out what they perceive are the company's business problems or what will improve their competitiveness. Either way you have information that will assist the positioning of your solution.

13. The other side's general business trends

Following industry trends is invaluable. After all, as we already mentioned, we sell in the context of the other side's weaknesses. To understand their weakness, we need to understand their relative position in their industry, their business initiatives, promises they've made to their customers and shareholders, and other similar information.

How do you get the information you need to be fully prepared for a negotiation? List some methods in the workbook.

STOP if you are using the Companion Workbook.

Exercise 8-2: Armchair detectives

GATHER YE ROSEBUDS

One of the best sources of information about the other side is—the other side. After all, who knows more about them then they do? Besides visiting their facilities, here's how we gather information from the other side.

Our mothers often give the best advice. Here's a sample: Nature gave you two ears and one mouth, so you should listen twice as much as you talk.

1. Ask open rather than closed questions

Questions that start with *how* and *why* will get you more information than questions that require a yes or no answer. Of course, sometimes you want to get an exact answer or put someone on the spot. If so, ask a yes or no question. This must be a conscious decision. Also, ask questions that lead to further discussion. Here are some examples:

Why have we been called in to bid?

- Why is there a need for our solution?
- How can we help you do business better and smarter?

- What do I need to know to help you find more effective solutions?

If you want information, prioritize your questions. If you want confirmation, ask questions for which you think you know the answers. Taking the conversation to unexpected paths can be very revealing.

To gather general information, ask all different kinds of open-ended questions. Of course, asking yes or no questions is also important to confirm what they are telling you. But remember, a "No" answer generally requires an open-ended follow-up, "Why?"

2. Listen, listen, listen

Do as your mother's proverb says. But this is easier said than done. Just think about how common poor listening is and how it can affect your credibility in the simplest situations. We talked earlier about flaws in listening. Let's use a common example of self-centered listening. As "Faline Renaldo" is introducing herself to you, what are you thinking about? She is saying "Hello, I am..." and you are thinking about how to introduce yourself. But you know your own name and shouldn't have to think about it. As a result, you don't hear Faline's name. Fifteen minutes later, at a break, you say to another member of her team, "As what's-her-name [referring to Faline] said..." Now the person you are speaking with is thinking, "What a bozo, Faline just introduced herself." Faline may even overhear this. She is thinking: "This person doesn't listen...I just introduced myself."

Your credibility is affected, isn't it?

How do you avoid this kind of scenario? How do you become a better listener? What if you repeat Faline's name when she introduces herself? "Nice to meet you, Faline," you say as you shake her hand. Since you repeated her name, you heard it twice and you have a much better chance of remembering it. Use the same method when you gather information. Repeat what you hear. "Let me make sure I understand…" Not only are you making sure you listened effectively, you are confirming that there is no misunderstanding. And usually after you repeat what you hear, you will be offered additional information to fill in any gaps. Now that's gathering information!

Repetition is key to effective listening.

3. Display a sincere interest in the person

You'll get more information by really caring, which you do. You can show your interest by making eye contact, leaning forward in your chair, and focusing on the speaker. However, in certain societies making eye contact is a sign of confrontation, so you have to be prepared by understanding something about the culture of the people you are negotiating with. More on this later.

Think about people who are in the professions of eliciting information from those who have a hard time relaying that information: psychiatrists, psychologists, members of the clergy, or maybe even bartenders. How do they get people to talk? They use words of understanding, empathy. They say things like "I see…" "I understand…" and "Please tell me more…" These phrases show interest and add to the comfort level the other person has in being able to relay information and being able to confide. Remember the empathetic listening we discussed earlier? These phrases generally avoid the use of the word "you," which tends to make people defensive and stifles the flow of information.

4. Observe body language

What is most telling about body language for most negotiators is when body language changes as a reaction to something that is said. Very often these changes are picked up by someone on the team, because the lead negotiator is busy. Here is another story:

> *I was working with a client's technical director. While I was engaged with the customer's lead negotiator, who was disagreeing with our value argument, their technical person (who earlier seemed disinterested) leaned forward and started shaking his head up and down in apparent agreement with us. During a break, my client's technical director pointed this out to me, so I asked if he could get this technical person to explain our position to their lead negotiator. He did, and they conceded the point.*

In certain cultures, body language can carry an entirely different meaning from what we expect. Mladen says:

> *I remember watching coverage of the Los Angeles riots years ago. A number of people were interviewed after some Korean-owned businesses were looted. Some onlookers said, "It's not surprising, since the Koreans are not friendly. They never look you in the eye." To many native Koreans (as in certain other cultures), making eye contact is a sign of disrespect or hostility, quite different from the U.S. culture, where eye contact is usually a sign of sincerity.*

Know and understand where you are and with whom you are dealing!

5. Pump up your comprehension

Start with your listening skills. As you learned in Chapter 3, listening is a skill that can be improved with practice. When you are eliciting information from a customer, client, or other source, you can improve your comprehension by:

- Repeating what you just heard

- Asking questions to clarify confusing aspects

- Confirming that the meaning of what you just heard is what was intended

If appropriate, you might also take notes. In most cases, we regard taking notes as a compliment. It says, "What you are saying is so important that I don't want to miss anything." These techniques show you really care about the topic and help you retain information you have gathered.

6. Set up informal meetings

Getting people out of the office often relaxes inhibitions and loosens tongues. People in a formal office setting tend to be on their guard. That's partly because they expect others to make judgments about what they say. The expectation—not necessarily the reality—is that informal settings are less threatening. Some informal settings to consider are:

- Restaurants

- Athletic or other spectator events

- Athletic activities—golf, sailing, basketball, racquetball, etc.

- Company or group outings

- Entertainment such as live theatrical performances

But most of our business is done in a more formal business setting. There are a few techniques you can use to create greater comfort and enable better information gathering even in an office setting. As mentioned, being empathetic and avoiding the use of the word "you" may create a situation of greater comfort in sharing information. Also, sharing some personal information often helps your counterpart relax and share information you need to know. Naturally, always stay within your comfort zone and your customer's.

In the end, you want to gather as much information as possible, given your time constraints. It's up to you to arrange your questions in order of importance and to listen to the answers.

Chapter 8: Negotiation Steps: Information Gathering

When asking questions, you may wish to avoid using the word "you" because it makes people feel threatened. Instead, use "we" or the company's name.

As mentioned, there is no information we don't want other than what we consciously do not want to receive. Even false information can be valuable because it clues you in to their motives and the power structure within their company. It can even give you leverage. Why would someone have false information? Are they not important enough to have the truth? Why would they pass on lies to you? Are they honest and think the information is true? Or are they trying to mislead you to serve their own interests?

Always remember M.O.R.E. from Principle #1 in Chapter 6. It is important to understand the motivations of the other company and the motivations of the individuals who represents or run the other company. Knowing what's important to the other side opens the doors of persuasion.

Here's the process:

Negotiation Steps "Information Gathering"

- **Who** do we want information about?

- **What** kinds of information do we want (and why)?

- **Where** are the sources of information?

- **How** (techniques) do we get information from our sources (including the other side)?

MOTIVATIONS AND OBJECTIVES

How do you go from mere communication to being believable and persuasive? You must understand the other side's motivations and objectives. Building value is your way of getting the customer to come to the decision you want. Build your case and you build value.

Chapter 8: Negotiation Steps: Information Gathering

WHAT YOU LEARNED IN THIS CHAPTER

- Information is key to a successful negotiation. Recall Principle #1: Get M.O.R.E.—Preparation is key to a winning negotiation.

- Get information about the company's organizational structure, decision makers within the company, the line of business, and anyone in the company who has influence over the potential outcome of the deal.

- Find out how decisions are going to be made in the company, internal office politics, external politics, how people are evaluated and measured, financial status and stability, timetable for purchasing, the company's history, and other facts.

- Sources of information include past contracts, fellow employees, trade shows, annual reports, company's web pages, general business reports, memorabilia in the client's office, venture capitalists, the employees, and personal observations.

- Techniques for gathering information include asking open questions, displaying a sincere interest, concentrating, holding informal meetings, listening, and repetition.

- Even false information can be valuable, provided you know it's false!

CHAPTER 9: NEGOTIATION STEPS:
INFORMATION MANAGEMENT

Question: When you are being introduced to someone, is it all right to say, "I've heard a lot about you"?

Answer: It depends on what you've heard.

Now, *that's* managing information!

In Chapter 8, you discovered how to gather the information you need to help prepare for a negotiation. In this chapter, we'll show you techniques for managing information. Start by completing the activity in the workbook: Read the scenarios and decide the appropriate action to take.

DAMAGE CONTROL

STOP if you are using the Companion Workbook.

Exercise 9-1: Walk this way

Read each scenario and the choices. Then select the choice you think best helps you manage information.

Scenario 1:

You are confronted by a tough issue—the customer wants to know the result of the latest testing on your new systems. The results are bad: The systems, which are supposed to be released for sale in 30 days, have been failing at a rate of 30%. This is far higher than is generally acceptable at this late stage for release of the product.

Scenario 2:

> *You are in negotiations with a company when they confront you with a competitive issue. They ask, "So, what happened to the negotiations you were having with my main competitor? I understand the negotiations have collapsed. Is that true?"*

Scenario 3:

> *During a negotiation, you create a solution to an inventory problem your OEM customer is having. After communicating it to the other side, a member of your team tells you that your solution can only be implemented manually, at a cost of $250,000. You made an honest mistake, but you would rather not increase your costs by $250,000.*

In each case. Choice #3 is probably the best.

Of course, your approach will depend on circumstances. For example, if the overall deal is worth $50 million, you may make a *conscious decision* to pay for the $250,000 mistake, but not before you tell the other side you are doing this for them. And try to make it a principled concession where you are getting something in return. After all, you are giving them at least $250,000 worth of services that have value.

Here's the reality:

During a negotiation...

1. You will be confronted with difficult issues regarding subjects you would rather not discuss. You can be truthful, but still manage information so you minimize the negative impact on your leverage.

2. People make mistakes.

3. People have trouble addressing mistakes or bad information. As a result, they often "sweep the problem under the rug" or ignore the problem rather than deal with it. That's human nature.

Chapter 9: Negotiation Steps: Information Management

Almost everything we do in a negotiation affects either credibility or leverage in some way—and usually both. That's why you must manage information thoughtfully.

Below are some proven ways that we manage information. These techniques should work for you as well.

MANAGEMENT TECHNIQUES

1. Admit mistakes sooner rather than later

You've learned from us that it's usually best to negotiate a deal on its merits. If you focus on the merits and confront mistakes quickly, you should be on firm ground. So, if you make a mistake, don't be afraid to admit it promptly. If you try to hide mistakes, more than likely they will be discovered and your credibility will be affected. Making mistakes damages your credibility a little. Hiding them for later discovery damages your integrity. People will have a hard time dealing with someone who has neither credibility nor integrity!

Is the other side going to give you a hard time if you make a mistake? Of course; they *should* try to erode your Negotiation Capital©. That's their job! But it's *your* job to maintain your integrity and credibility. You are always trying to build credibility to gain leverage and to build a long-term relationship with the customer. Remember Principle #5 from Chapter 6: "Negotiation is a continuous process."

Once again, we mention this concept of Negotiation Capital. It's a term we coined to describe the amount of willingness by either side to negotiate and exhibit flexibility such as making trade-offs and concessions. When you lose credibility, you erode your Negotiation Capital.

Think of it this way: Negotiation Capital is like currency. If you have a good relationship and have done a good job persuading the other side (creating a perception of leverage), you have earned Negotiation Capital. This "currency" translates into the other side's willingness to move closer to your way of thinking. If you lose credibility or have to backtrack due to mistakes, then you use Negotiation Capital as if you were "burning" currency. The other side's willingness to deal with you and to compromise is reduced. This gets exacerbated when they also lose faith in your integrity. You will have to work that much harder to earn back Negotiation Capital, or you may wind up without a deal. Your Negotiation Capital is manifested in terms that the other side is willing to give and alternatives they are willing to explore.

Long ago, the Greek philosopher Aristotle explained what a person needs to be successful. In *Rhetoric,* Aristotle wrote that people need more than an ample vocabulary and good taste to pick the right words for the right occasion. They need more than intelligence, self-control, and balance. They even need more than being up-to-date on the issues— although all these accomplishments and traits are useful.

Above all else, Aristotle claimed, good speakers also have to be good people. If you want a group of people to accept your ideas, you must be respected and trusted. People don't listen just to your words; they also focus on you. To make a value argument, you need to be respected by others and by yourself. Taking ownership for any errors that may occur is a good way to build integrity.

Harvey recounted this story of taking responsibility for one's mistakes:

> *I was negotiating a deal when I got a call from the lead negotiator on the other side. We'll call him Bert. He shouted, "Harvey, your financial guy is trying to cheat us. The prices he gave us are incorrect. The prices should be lower for the system configurations we requested. You are*

way overpriced. And the guy on your team who does your pricing, that Charlie Barnes? He's a liar!"

I let Bert rant until he ran out of steam. Then I calmly said, "You mean that you believe Charlie made a mistake. I'll review it and get back to you." At that time, Charlie had just joined my team. He was a studious guy, a very sharp Ivy League graduate. He went on to have a brilliant career, rising to vice-president in the company. Charlie was an honest man, too. I reviewed the numbers with Charlie and we both realized that he had indeed made a mistake. I called Bert back.

I said, "Congratulations, you were right—this was a mistake."

Bert replied, "Charlie is a liar and I never want to see him again."

I said, "First of all, this is a team mistake. I should have looked at the numbers more closely. Also, I saw the expression on Charlie's face when we figured it out. He was stunned. But everyone makes mistakes. And Charlie has a great track record, impeccable honesty, and will continue as part of my team. Let me talk to you privately next week when we come to see you to resume negotiations. I hope that you'll be there. Our team, including Charlie, will be there also."

Bert was in the room during our next negotiation and everything was cool. He realized that as lead negotiator, I had taken responsibility for the mistake someone on my team had made. Bert had tried to convert an honest mistake into an integrity issue to gain leverage. It didn't work. We maintained our credibility and our team while getting the deal done successfully.

The K&R Deal Forensic

This scenario provides several important lessons:

1. Mistakes happen.
2. By quickly confronting mistakes we make, we can maintain credibility and integrity.
3. By supporting our teammates, we can prevent the other side's divide-and-conquer tactics, now and for the future.
4. Teamwork is key to successful negotiations. Part of teamwork is good team management and support, even when a mistake is made.

If you confront mistakes or other issues head on, you can address them in a manner and within the timeframe that gives you control rather than in a reactive mode. You can:

- Steal the other side's thunder and increase your leverage

- Reduce their anger and indignation

- Reduce your use of precious Negotiation Capital

2. Wait

Recall what you learned about patience and listening in Chapter 4:

When you negotiate, take your time. While time is always an issue, rarely does the deal have to be done at that moment (yes, even at the end of the quarter). Often, you can manage information far better by being patient and letting the other person talk. This enables you to hear information you may be able to use to make your value argument, or manage your weaknesses.

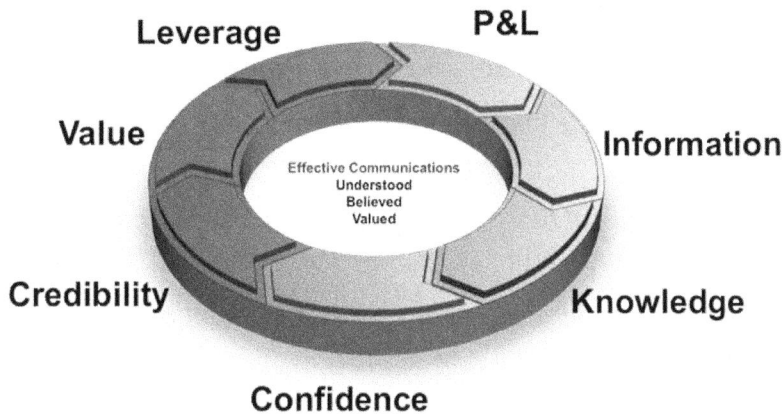

Figure 1: Leverage Cycle™

3. Promise less, deliver more

Of course, never promise what you can't deliver. Mladen relates the following example of promising less but delivering more.

> *I was doing a deal with Newest, a young technology company that wanted to expand into Asia and Europe. The company I was representing was a giant in the telecommunications industry worldwide. We could give Newest a big advantage by issuing a press release and advertising the relationship to our distribution network in Asia and Europe. Newest's negotiator understood the leverage issue and the importance of linking their name to ours, but did not want to acknowledge it for fear it would cost him money. "No," he said, "your advertising is of no value to us." This was a ploy to devalue our offer, as we had already written a term about the press release and advertising into the contract.*

> *Rather than completely deleting this term, we just put a line through it, so they would continue to see it in all subsequent drafts of the agreement. At the end of the negotiation, Newest's negotiator said, "I see the press release and advertising provisions are still crossed out…"*

"You're right," I replied and dropped the issue. We closed our discussions for the day and I went to see my client.

She said, "Great job! Let's get the deal signed now."

I replied, "Give me another week; we can get $1 million more."

I was confident that putting the press release back in would get them to give us $2 million more, since giving them international exposure had a potential annual return of $10+ million in the first 18 months.

The next day, their negotiator said, "Well, a joint press release could help us get recognition."

"Sure would," I replied. By the end of the week, we were able to use the press release to get another $2 million for our team.

And it would definitely help Newest; it had tangible value, so why would we give it away?

The K&R Deal Forensic

Lessons learned:
1. Experienced negotiators will try to mask terms that reduce their leverage.
2. By understanding the other side's posture in the marketplace, we were able to recognize the unique value of certain terms.
3. Persistence in withholding a valuable term forced the other side to acknowledge it value.
4. With acknowledgment of the value of a term comes a willingness to pay.
5. Leverage was used wisely to satisfy both sides.

---

4. Recognize that changing people is hard work

Some people will understand the value argument you are building. They are rational, intelligent people who understand how leverage is used. They can be a joy to negotiate with because they know how to deal and they enjoy the process.

However, negotiating with other people can sometimes be more difficult. Adversarial negotiators, for example, may resist all rational arguments. With these kinds of negotiations, it is most important to be persistent and focused on the merits of the deal. Alternatively, uninformed or unprepared negotiators may resist rational arguments because they are afraid to make decisions. It is the fear of being wrong. The only way to get them turned around is to be patient and train them through repetition of rational arguments or discussions with others on their team who may know more. Otherwise, your only alternative is capitulation or termination of discussions.

In addition to management techniques, you can use blocking techniques as you negotiate. Always consider your purpose, audience, and personal style as you decide how, when, and if you want to use these strategies.

Negotiations are not one-sided. They are an exchange of ideas and information. Just as you are trying to find information from the other side that gives you leverage, so they are trying to gather information from you. And when they ask difficult questions or ones that you don't want to answer, what do you do? Negotiators use blocking techniques that come naturally to most people.

BLOCKING TECHNIQUES

1. Change the subject

If you don't want to answer a question, just change the topic.

Your response doesn't have to be as obvious as "How about those Olympics?" or "Isn't this some fine weather we're having?" Instead, you may subtly shift to another item on the agenda or another aspect of the negotiation. Often, changing the subject means going back to a more comfortable subject discussed earlier. You can use "As we had discussed..." as a segue into the previous topic.

2. Give a broader or more narrow answer

Think about politicians. They often block information by being general rather than specific. By the time the inquirer realizes that the answer lacks detail, they will have moved on to another topic.

3. Answer a question with a question

Mladen suggests this blocking technique because it can help you elicit more information. That's because you are turning their fact finding into yours. For example:

> Customer: "What is the capacity of that plant you are building in Malaysia?"

> You: "Hmm, that's a very interesting question. Why do you ask?"

So, then you are gathering information regarding their requirements and concerns. This gives you the opportunity to tailor your value statement to their stated aims.

4. Use body language

As mentioned in Chapter 8, this is not an exact science, but you can send strong blocking signals with body language. For example:

• Crossing your arms across your chest

• Placing your hands on your hips

Remember: Body language can be culturally based, so don't assume that a customer is reading your body language the way you intend.

Be careful reading other people's body language if you are not an expert. For example, a person can have their arms crossed because they are comfortable, while someone might regard their position as hostile or confrontational.

Chapter 9: Negotiation Steps: Information Management

Complete the activity in the workbook to see how well you read body language.

STOP if you are using the Companion Workbook.

Exercise 9-2: Body wars

5. Use humor

We heard this great story:

> *Lucy, a sales manager, stood before a group of key customers who had been gathered to observe a demonstration of her company's state-of-the-art technology.*
>
> *The monitor blurred and the links wouldn't work. Lucy's attempts to address the issue were futile. Lucy called her company's tech support for assistance, but the liaison was gone for the day. She took a deep breath, faced the group, and said: "This concludes my demonstration of our competitor's product. Next week I'll come back and show you ours."*

Lucy was quick on her feet and the laughs she got helped temper the effect of the key question: Why doesn't your system work? As you will learn in a later chapter, only use humor if you're comfortable with it and the environment.

Use blocking techniques carefully and recognize that using blocking techniques too frequently will eventually cost you credibility.

Complete the workbook activity about effective persuasion techniques.

STOP if you are using the Companion Workbook.

Exercise 9-3: Block and tackle

In the previous workbook exercise the company was using persistence. Repeating your argument is one of the most effective negotiating tools you can use. Read on to find out why.

Remember: While one side is persistent in its requests, the other side can be equally persistent and consistent in its response.

INFORMATION UTILIZATION/PERSUASION TECHNIQUES

1. Repeat the argument; be persistent

Ever been a part of this negotiation?

09:00	Child:	"Can I have a cookie now?"
	Parent:	"No, Bobby, it's too early in the morning for a cookie."
09:03	Child:	"Can I have a cookie now?"
	Parent:	"No, it's too much sugar."
09:07	Child:	"Can I have a cookie now?"
	Parent:	"Cookies are bad for your teeth. Please stop."
09:09	Child:	"Can I have a cookie now?"
	Parent:	"You're making me crazy."
09:10	Child:	"Can I have a cookie now?"
	Parent:	"Here's the cookie. Have another. Have the whole package."

Children are brilliant negotiators. Why? Because they are *persistent.* They also have no sense of time pressure. As a result, they repeat the argument over and over. They often get their way because their persistence outweighs their parents' patience.

As adults, we have less time and constant time pressure, so we are less persistent. Yet our arguments are much more complex than children's arguments. The first time we make an argument, the other side may struggle just to understand it. And they might not be good listeners. So what right do we have to expect people to agree with us if we don't make our argument again? Repeating our points is a great way to make our case. What happens? After hearing an argument for the third or fourth time, the person on the other side can usually repeat that argument. More often than not, they will do so. Even if someone disagrees with us, they will often repeat our arguments inside their company just to illustrate why we are taking our position and to explain their own point. In doing so, they inadvertently become advocates for our position.

Mladen suggests that you vary your words and intonation as you repeat your argument so you don't sound like a broken record, but still make your point.

Axiom: Repetition is key to effective persuasion.

We discussed earlier that you often sell in the context of customer weaknesses. In fact, deal making in general is all about solving the other side's weaknesses. But be careful. Extreme organizational pain usually causes corporate upheaval, resulting in "paralysis of indecision". Even if the pain is dictated by a critical weakness and you have a solution, it usually takes extra persistence to make the deal.

The following scenario comes from Mladen's files. It occurred near year-end:

> *A software company we'll call "Eager," with a sales staff of approximately twenty, was engaged in discussions with a major worldwide financial institution, "Global". In the middle of the third quarter, Global had laid off thousands of employees, citing millions in monthly losses. The losses were largely due to the publicized difficulties Global had in processing Internet and e-commerce transactions. At the*

same time, all budgets, including new technology purchases, were frozen.

Eager had a unique solution that would remedy Global's problem within weeks of implementation. Eager's negotiation team built a compelling business case and ROI, and set up meetings within one of Global's business units most affected by the circumstances. A group of Global's managers (all new in their jobs due to the reorganization) reviewed Eager's demo, and all agreed that this solution would solve their problem. However, they said they couldn't make a purchase decision and that Eager should talk with another group of operations managers.

Eager did so, and the other operations managers also agreed that the solution would solve their operational problem. But they couldn't make a decision either. They sent the Eager sales team to yet another group.

This process repeated itself several times. Each group agreed that Global needed Eager's solution, but none would make the buying decision because of the risk in their volatile environment. Meanwhile, a new CIO had been appointed by Global who was too busy to attend any of the meetings.

It was now late November. Frustrated, the Eager sales team was about to give up on their largest deal of the year. To her credit, Eager's vice president of sales said, "We have too much invested in this effort and the solution makes too much sense to give up on the deal." She called on Global's new CIO directly and asked for thirty minutes of his time.

They met in early December.

After fifteen minutes of the demo, the CIO stopped the business case presentation and said, "Why haven't I been told about this before? This is what we need. I want this implemented as soon as possible!" Budget dollars for a partial rollout became available, with a full implementation scheduled for the first part of the following year. The contract was signed before Christmas!

The K&R Deal Forensic

At least four factors came together in the successful close:

1. Satisfying a critical need with a unique solution supported by quantified value (business case) often does sell.
2. Recognizing that "paralysis of indecision" associated with a reorganization can be overcome with persistence (rather than throwing discounts at a problem).
3. Getting to the right decision maker whose measurements are personally affected by the solution is crucial.
4. Many companies spend money, even when budgets are technically "frozen." You just must be creative and compelling enough to invoke a thaw.

---

2. Inducements and incentives

Consider illustrating incentives to give the customer a reason to close the deal. We are not talking illegal gratuities here. Legitimate inducements can take many forms, depending on your product or services offering. But never forget what you learned in Chapter 6, Principle #4: "Concessions easily given appear of little value." The inducements should be tied to your value argument to offer real value to the customer. That means we don't favor inducements that are "gifts" to the customer. These would be classified as *unprincipled* concessions. That costs you, and usually doesn't buy you a whole lot, if it's perceived as an easy concession.

Rather, in many negotiations, motivation to make decisions by either side is based on a positive or negative inducement that repeats the value proposition. Here are four examples:

(a) Positive inducement:	"Look at the positive impact this deal has on your business! The sooner you make a decision, every week you stand to gain $XYZ."
(b) Positive inducement:	"Based on our discussions this deal should reduce your infrastructure costs by 3% and if you sign now you save 5%."
(c) Negative inducement:	"If you don't make this deal, this is what your business will lose."
(d) Negative inducement:	"If you don't sign now, you will continue to have a non-competitive cost base and lose the 5% bonus."

Note that all of the above inducements need to be quantified with rationale that is accepted by the other side. That agreement is what makes it a value statement, regardless of whether it's positive or negative. In fact, the best inducements are rooted in a business case.

It's interesting that almost all inducements can be put either positively or negatively, yet most of us don't think about how we should phrase these statements. We need to understand that people perceive positive and negative inducements differently. In most cases, skilled negotiators use positive inducements because most people receive them better.

However, negative inducements can also be used to great advantage. In fact, research presented in George Kohlrieser's book, *Hostage at the Table,* shows that people act with more urgency in relation to risk and negative inducements. You can use negative inducements deliberately to create worry. But be careful when using inducements. If the inducement contains an implied threat that you don't carry through, you lose credibility. In fact, we generally do not favor inducements *b* and *d*, unless the 5% is justified by business rationale and comes off the table if the deal is not signed as requested. And before such an inducement is made, the value has to justify the deal!

Inducements need to be thoughtful. Ask yourself, "Does this inducement motivate the right behavior?"

3. Ask more questions

In addition to repeating the argument, you can often effectively manage information by asking more questions. This is related to the blocking technique of answering a question with a question. This technique helps you elicit the information you need to be sure you understand what the other side wants and needs. In the next chapter, you'll use K&R's MID™ to learn how to distinguish among mandatory, important, and desirable goals, issues and requests. Asking questions can help you do this.

4. Rephrase the question

Ever hear this joke?

> *A grasshopper walks into a bar and the bartender says, "Hey, we have a drink named after you."*
>
> *The grasshopper says, "You got a drink named Bob?"*

Sometimes it's an issue of vocabulary. Rephrasing the question in your own words can help you make sure you understand what the other side is really asking. Rephrasing the question helps you focus and therefore listen more carefully, too.

5. Work as a team

In Chapter 6, you learned Principle #3: "A team divided is a costly team." One of the best ways to utilize information is to work with your teammates. You'll learn more about the importance of teamwork in

Chapter 15.

6. Choose a site carefully

You know the three rules of real estate: "Location, location, location." The site of a negotiation *does* matter. Here's one of our concerns regarding telephone negotiations:

> *People often want to stay in their own office. Why? So, they can keep on working while they try to carry on discussions. Notice that word "try". Yes, we're all far too busy, but holding a negotiation by telephone usually creates challenges in its own right. Having all the participants in the same room helps us communicate much better. This way, everyone is giving the negotiation their full attention. In addition, you can manage the dialogue better and prevent people on your own team from misspeaking.*

You can manage and use information much more easily if you are all in the same place, negotiating person to person. But negotiations by conference call are a part of life. There will be more on telephone negotiations later in Chapter 14.

As we have said: "If you're going to invest time and resources in a deal at some inconvenience, the other side should do the same. So if the other team demands that you fly across the country to their home base for a negotiation this week, let them fly across the country to your home base for the next negotiation session. Of course, the playing field is never level, but requiring the other side to invest time, money, effort and resources helps validate their commitment and raise their psychological stake in getting the deal done with you."

Let's sum it up in visual form:

Negotiation Steps: Information Management

- Management Techniques—Dealing with information about yourself

- Blocking Techniques—Responding to their inquiries

- Information Utilization/Persuasion Techniques—Steps to help them agree with you

NEGOTIATION STEPS: STRATEGIC PLANNING

Strategic planning starts with a business plan. To implement that plan, you must consider strategic and tactical acts. In general, use tactical acts to support strategic ends. The following chart summarizes these two types of acts.

Business Plan: Tactical Acts to Support Strategy

	Success	**Failure**
Strategic Act	Directly responsible for success of strategy	Direct impact: may need to change or eliminate strategy
Tactical Act (or interim solution)	Can help support strategy, but at least "holds the line"	May be costly; but should not directly impact or alter strategy

But tactical decisions do have long-term strategic effects. For example, once you make a discount, the customer knows you can offer it again. To look at this issue in more depth, complete the activity in the workbook.

Even "tactical" acts require strategic planning.

Chapter 9: Negotiation Steps: Information Management

STOP if you are using the Companion Workbook.

Exercise 9-4: The king of tablets

Consider the following scenario:

> *Fred's business plan is to become "king" of Major Brands tablets for all of Cook County. He obtains financing based on the plan and gets a deal from Major Brands for 2,000 tablets to be shipped to him within two to three weeks. To that end, Fred calls his contractor about getting into his nice new store. the contractor says, "Fred, unfortunately your store cannot be ready for another six months due to drainage problems on the property." Fred clearly has a problem: What can he do with all those tablets for six months? After all, he has to start making sales to pay the creditors. Then Fred remembers his friend Phil, who owns Phil's Electronics.*
>
> *Fred calls Phil. Phil says, "No problem, Fred, old buddy. I'll be glad to sell your tablets through my outlet until you're ready to open your store." They negotiate a deal: Phil gets a 35% cut of sales. The next day there are billboards all over town. "Phil's Electronics now offering Major Brands tablets!" Phil is a huge success. He sells out and orders more tablets from Major Brands. He becomes the "king" of Major Brands tablets. Phil has established his name as the Major Brands tablet supplier for all of Cook County.*
>
> *Six months later, Fred's store is ready. But Fred has a problem. Was the deal Fred made with Phil successful? Was the deal tactical or strategic in its intent? In its execution?*

Let's take the second question first. Fred clearly intended the deal with Phil to be a short-term interim solution to a problem. So, with respect to the first question, as a tactical act, it was somewhat successful. Fred is able to pay off his creditors. But the execution of the deal caused his strategy to be destroyed.

Chapter 9: Negotiation Steps: Information Management

What was it about this deal that made it a strategic disaster for Fred? He didn't set the right expectations and focus on his objectives: to be the "king" of Major Brands tablets for all of Cook County. For one, Fred's name and not Phil's should have been associated with Major Brand's tablets in Cook County. The contract with Phil should have addressed how the tablets would be marketed by Phil. These should have been *Fred's* tablets available through Phil's Electronics, not *Phil's*. Fred should have negotiated some limited non-compete agreement with Phil, and remained the contact with Major Brands. (Logically, he would have tried to get an exclusive from Major Brands, but the supplier probably wouldn't do that for an unproven market entrant.) That way he would have some control over the chain of supply, and therefore his strategy to be the "king". Remember M.O.R.E.? The requirements (the terms) must relate to the long-term strategic objectives; otherwise, even a successful tactical act can have devastating long-term strategic consequences.

WHAT YOU LEARNED IN THIS CHAPTER

- Almost everything we do in a negotiation affects either **credibility** or **leverage**—and usually both. Successful negotiations include managing information.

- **Management Techniques** include admitting mistakes sooner rather than later, offering inducements, promising less and delivering more, recognizing that changing the way people negotiate is hard work, and keeping confidentiality.

- **Blocking Techniques** include changing the subject, giving a broader or more narrow answer, answering a question with a question, using body language, and using humor. Be careful - abusing blocking techniques can lead to credibility issues.

- **Information Utilization/Persuasion Techniques** include repeating the argument, asking more questions, rephrasing the question, having patience, reading body language, working as a team, and choosing a site carefully.

- Even in a tactical deal, the requirements (the terms) have to relate to the long-term strategic objective or you may jeopardize your strategy.

CHAPTER 10: K&R'S MID™

"I MUST have a ten-day shipment guarantee!"

"On-site system support is a deal breaker!"

"This functionality is a must!"

"Listen to me—a price reduction is mandatory."

"It is imperative that we control service changes!"

How often have you heard demands like these during a negotiation? We bet you've heard statements like these far too often! It's as if every one of these requests must be met or no deal! People have trouble arranging the issues in order of importance during negotiations. They often speak of everything as "mandatory", "must have", "need", and "deal breaker". The use of such terms paints people into a corner. But, in reality, there are very few deal breakers in negotiations. In this chapter, we'll help you recognize those deal breakers and prioritize issues through K&R's MID™ Chart (or simply MID). One of the goals of this book is to enable you to break down the issues so you can identify true mandatory goals or the "ends" and reduce conflict over the means. Understanding the true goals or ends helps us close deals more easily.

CONFLICTING GOALS

Let's start with a page from the life of our friend and colleague, Jim. Complete the activity in the workbook to decide how you would solve the problem Jim faced after a long day at work.

STOP if you are using the Companion Workbook.

Exercise 10-1: Apples of my eye?

Our colleague, Jim, had three young children. To save them needless embarrassment, we'll call them Child 1, Child 2 and Child 3. Each child wanted an apple; but due to the demands of school lunches, snacks and other healthful eating, there was only one apple left. And it was too late to go to the market to buy more apples. Here was the dialogue:

Child 1 says: "I MUST have the apple."

Child 2 says: "I NEED the apple."

Child 3 says: "Let me have the apple."

Jim's wife turned to him and said, "You're the big-shot negotiator, you work this out."

You could offer to cut the apple into equal pieces and give each child a third. While that seems like a "fair" solution, how do we know it will satisfy any of the children? It may be a truly fair solution if the reason each child wanted the apple is to eat it. That way, each gets their fair share. But how does Jim know what they want the apple for? What problem is Jim trying to solve? What problems are the children trying to solve? What are they really after? These are the kinds of questions Jim needs to ask before he can offer a fair solution.

It turns out that each child needs the apple for a different purpose.

- Child 1 says: "I MUST have the apple. I need the peel to make model huts for my social studies project. The peel is perfect for the canopies on the huts."

- Child 2 says: "I NEED the apple. I need the seeds for my science project. I will grow apple seedlings for the science fair."

- Child 3 says: "Please, let ME have the apple. I need to eat an apple every day for the vitamins...remember, *an apple a day...*"

Now, Jim knew how to solve the problem. And rather than having to compromise and get a third of the apple, each child's purpose for wanting the apple in the first place can be satisfied 100%. That's because Jim understands what questions to ask to separate the means (the request each child makes for the apple) from the ends (the purpose for which each child feels a need for an apple). And now you will, too. The key question is: What problem are we really trying to solve?

MEANS VERSUS ENDS

The means are alternate ways of accomplishing goals. The ends are the goals. You can also distinguish between means and ends this way:

Means = ways of getting there

Ends = there

People naturally talk in terms of means; that is, they articulate the easiest way of getting what they need. For example, if you are thirsty and prefer water, you are more likely to say, "Can I have a glass of water?" than "I am thirsty and want to relieve my thirst. What do you have to drink?"

That's because you have already solved the problem in the way that you prefer. But what if the other person does not have water? Then they have no way of satisfying your specific request. But if they know you are thirsty, they may have all kinds of beverages to quench your thirst. Separating the means from the ends is crucial to understanding what people really want and need. Then you can identify creative alternatives to satisfying their needs. That's where K&R's MID comes into play.

SEPARATING AND UNDERSTANDING MEANS AND ENDS: INTRODUCING K&R'S MID™

MID stands for the three main types of requests, which may or may not be actual goals. Here's the key:

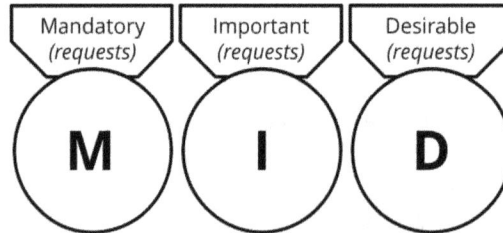

Figure 1: MID Definition

- *Mandatory requests* should be the true goals (or ends) that must be achieved by either side for the deal to work. These are the true deal breakers. They might concern price, but they are equally likely to concern any other variable terms in the negotiation such as delivery dates or tech support. Usually only ends are mandatory. The exception would be when there is only one way (means) to accomplish a mandatory end. To clarify, true goals equal ends, and we use the terms interchangeably.

- *Important requests* are goals or favored means that matter a great deal to the other side or to you. But the success or failure of the deal is not likely to depend on them. They can be negotiated for other equal or more important concessions.

- *Desirable requests* are items on either side's wish list, but they have the lowest priority. It would be nice to give all parties everything they desire in a deal, but important and mandatory requests take clear precedence over desirable ones. Your "desirables" can be traded off to gain something of greater importance to you. That's why your "desirables" can be used as bargaining chips in negotiations. That's particularly true if what you desire is something that's conflicting and more important to them. Then you can trade it off for something more important to you, making a principled concession.

- *Conflicting requests* are items that are in direct conflict. The simplest example is price. If I can't pay more than $20 and you can't (or won't) sell for less than $22, we have a conflict.

- *Independent requests* are items that one side cares about and the other doesn't. For example, one side may be interested in doing a deal to help set a *de facto* standard in a particular segment of the industry while the other side may not care which standard prevails as long as they get a good deal.

- *Joint requests* are items that both sides should care about. That doesn't mean they care about them equally, but they care the same way. For example, it may be in the best interests of both parties for one side to use oil-based paint on the moldings. To homeowners, it is desirable since they like to easily wipe children's stains off the moldings. For the painter, the painting is done in one coat, which saves time to get onto the next job.

It is crucial that you are able to sort the requests into their categories. Doing that enables you to do a better job of meeting each side's needs, because you are distinguishing between the way people do business (means) and the goals they must achieve (ends).

K&R's MID Chart of Goals		Mandatory (Ends)	Important (Preferred means or ends)	Desirable (Desirable means, some ends)
Conflicting	Buyer			
	Seller			
Independent	Buyer			
	Seller			
Joint				

You can see that the chart arranges the requests from *most* to *least* important. The ranking works like this:

Chapter 10: K&R'S MID™

Most Important ➡ Significant ➡ Useful

Mandatory　　　　　**Important**　　　　　**Desirable**

Using K&R's MID can help you:

1. Separate the means from the ends.

2. Arrange negotiating issues in order of importance. This enables you to distinguish issues that are real deal breakers (mandatory goals) from those that are important or just desirable.

3. Move conflict over means out of the Mandatory column and into the Important or Desirable column, since means are rarely mandatory. Or, better yet, take conflicting means and, by figuring out the ends, eliminate the conflict. That's because the ends may actually not conflict at all (they are independent or even joint). There may be other means to satisfy non-conflicting ends. However, as long as the conflict remains in the Mandatory column, the deal shouldn't close.

4. Understand where the trade-off opportunities exist in the transaction.

   Not all deals can be made—or should be made. The MID analysis forces you to determine what problem you are really trying to solve. This is especially crucial if you have a conflict over goals that you really can't solve. If this is the case, the MID analysis can help you know when to cut your losses and move on.

USING K&R'S MID™

For each request either side makes, especially the ones that are conflicting, ask yourself the following two questions as you start to separate the means from the ends:

1. Why is that request being made?

2. "What problem are we trying to solve?" is always a key question. (Remember the apple story. Each child was making the same request. Those requests made it look like a mandatory conflict. Yet each child wanted to address a different problem, none of which were in conflict.)

To truly understand what you and the other side have as your goals/objectives, ask questions that are likely to elicit the answers. Then be patient in listening and understanding the answers. As you learned in Chapter 3, patience and listening are two keys to success.

Listen to gain understanding, not to argue.

Charting goals on the MID often involves placing requests in the form of issues and/or terms of a deal in different categories. Those issues may be means or they may be ends. If they are conflicting, always ask the above two questions to separate the means from the ends, and re-categorize the answers. You may find that the final ends are not conflicting. Read the following example to explore this topic further.

STOP if you are using the Companion Workbook.

Exercise 10-2: The magnificent conflict with BIG

Magnificent Inventory Management, Inc. (MIMI) is in the middle of negotiations with BIG, Inc. MIMI's inventory management software package is known industry-wide as the best inventory management suite on the market. BIG wants to license the MIMI package from MIMI for use with its public "Cloud" offerings for retailers around the world.

BIG's chief negotiator, Luther Large, has made it clear to the CEO of MIMI that he values their software because it's easy to install and integrate with BIG's systems. To that, MIMI's CEO replies, "This is a great match. Out software is ready for you with little or no modifications, and we have agreed to do the modifications you requested." The parties have agreed that MIMI will perform all maintenance and support of the package for BIG's customers.

Near the close of the most recent round of negotiations, after all the issues seemed to be ironed out, Luther says to the MIMI CEO, "Of course, you understand that we must have access to all your proprietary source code for the MIMI inventory management software package." MIMI's CEO replies, "Luther, surely you jest. The source code is our crown jewel. You know I can't give you the source code!"

Luther responds, "We have this relationship with all our licensors. If you don't give us the source code, we have a problem. A BIG problem..."

You have been called in to resolve this BIG problem. Use the MID as you answer the questions that follow and formulate your negotiation strategy. Note that we added some parentheticals under the "Mandatory," "Important," and "Desirable" headings. That is because the truly mandatory goals (or ends) are usually deal breakers. Particular means may be desirable or preferred, but it is the end goals that most people really care about. Of course, that is not always the way people speak.

Here's how we approach the problem of means (alternative ways to achieve the goals) versus ends (goals), as we discussed at the beginning of this chapter. Let's start with answering the questions we asked you to answer.

1. Is Luther using words that sound like a mandatory goal? We explain it like this:

   Luther is indeed using words that sound like a mandatory goal because he says, "I MUST HAVE access to your source. If I don't get it, I won't do the deal." This is what makes his demand seem mandatory.

2. Is MIMI's CEO using words that sound like a mandatory goal? Explain your answer.

Yes, MIMI's CEO is using words that sound like a mandatory goal because she says, "I CAN'T GIVE you the source code. The source code is our crown jewel." If you plot these initial statements on the MID, it looks something like this:

K&R's MID Chart of Goals		Mandatory (Ends)	Important (Preferred means or ends)	Desirable (Desirable means, some ends)
Conflicting	BIG (Licensee)	**"We must have access." (M or E?)**		
	MIMI (Licensor)	**"I can't give you that." (M or E?)**		
Independent	BIG (Licensee)			
	MIMI (Licensor)			
Joint				

3. If this is a conflict over mandatory goals, what happens to this deal?

Note that the first entries in the upper left-hand comer of the MID are the positions articulated by both sides. If they are truly mandatory ends (goals) that are in conflict, then this deal should not close. So, some analysis is necessary to see if these statements reflect irreconcilable mandatory goals or describe means of accomplishing different goals that are not conflicting or are either important or desirable. If they are important or desirable, they should not be deal breakers.

4. How can you find out whether this is really a conflict over mandatory goals (ends) or an argument over means?

    If positions are articulated as mandatory conflicts, until you get underneath the reasons for the positioning of each party and figure out the problem, the problem cannot be resolved. Recall the problem with thirst: When you are thirsty and you prefer water, you are more likely to say, "Can I have a glass of water?" than, "I want to relieve my thirst."

    So, in this example, water was a means, but you never articulated the problem, which was thirst. If I don't have water, I can't satisfy your literal request for water. But, if I know your end (goal), that you want to quench your thirst, I can offer many alternative solutions (means), even if I don't have water (i.e., juice, milk, soda, beer, tea). Each side should ask the other, "What problem are you trying to solve with your request?" Or, more specifically, "Luther, can you please tell me why you need the source code?" and, "What is it that MIMI is worried about? Why is it that MIMI can't give us the source code?"

5. If the two sides' statements reflect means, what are the likely ends (goals)?

    Let's deal with BIG's request first. Luther Large speaks as if this is a mandatory end ("we must have"), but it is likely that it's only a means. As mentioned, if it is an end, then it may be a deal breaker.

    If Luther's request is a means, BIG's goal may be to protect its customer to whom they, in turn, sell access as part of their cloud offerings. What would happen if MIMI, a small company, decides not to support the code in the future? Or, what happens if MIMI decides not to release any updated versions of the code to improve it, even though customers expect certain improvements? What happens if MIMI has financial problems or goes bankrupt? Finally, what happens if MIMI gets bought by someone else who does not want to do business with Luther Large's company? These are all significant issues for Luther Large. Getting access to source code today may be a means to solving these future issues, but it probably is not the only means. The end (goal) for Luther to support BIG's customers is closely related to something that affects BIG directly—its reputation. After all, it's Luther's and BIG's reputations that are at stake with these customers. BIG does not actually need the source code today because MIMI is fulfilling BIG's business needs.

K&R's MID Chart of Goals		Mandatory (Ends)	Important (Preferred means or ends)	Desirable (Desirable means, some ends)
Conflicting	BIG (Licensee)	**"We must have access."** (M or E?)		
	MIMI (Licensor)	**"I can't give you that."** (M or E?)		
Independent	BIG (Licensee)	Protect reputation/ company		
	MIMI (Licensor)	Protect key assets/ company		
Joint				

BIG might need the code sometime in the future or they might not. Only time will tell on that issue. The discussion about their potential relationship is the best time for Luther to protect his company for the future.

What about the Licensor, MIMI? The CEO has also taken a mandatory-sounding position: "I can't give you the source code." The MIMI CEO is probably concerned about protecting her most important company asset. If that asset is lost or devalued, so is her company. All companies (particularly technology or entertainment companies) worry about their intellectual property assets, which can be their lifeblood. For example, writers protect their intellectual property by copyrighting their books, articles, screenplays and other published works. This prevents others from using their works without giving credit and appropriate compensation. It's the same with a software company and their source code.

If we look at the MID, we see that the mandatory conflict has shifted, and we are really dealing with independent goals.

6. Are likely goals mandatory and conflicting?

The likely goals are for each company to protect itself: (a) BIG wants to protect its customers and thereby its reputation (its goodwill) and (b) MIMI wants to protect its key assets and its company. These goals are independent and do not seem in direct conflict. In fact, these are reasonable, rational goals for any company!

Once you know the true ends, any vehicle that gets you to those ends is the means. Once you know the true ends, you can be very creative about the means to accomplish them. Understanding ends usually opens up a portfolio of alternatives to solving problems.

7. Are there other means that can be used to accomplish these goals? If so, What are they?

Chapter 10: K&R'S MID™

Let's restate the problems to figure out some ways to solve them.

Here's the problem that Luther Large faces: "I'm worried that I am going to have thousands of cloud customers. If I can't provide continuous support, they won't do business with me. As long as MIMI is supporting my code, I'm OK. The problem begins when the MIMI company stops supporting the code for whatever reason. That is when I need to get access to the code, but I want to negotiate the issue now, not later when I am losing customers due to lack of support. It will be too late then!"

Luther also knows that waiting to negotiate this issue later when BIG is dependent on the MIMI software will make such a negotiation much more difficult for BIG. MIMI's problem with giving BIG access to the code is: "As long as I am a viable company, this is my key asset and I have to protect it. If I give you the source code, you could misuse it or use your vast resources to rapidly enhance it, making MIMI irrelevant in the market. That would be an irresponsible thing for me to do."

One potential solution to both parties' concerns may be for MIMI to license the source code to BIG today, with restrictions. From MIMI's perspective, these restrictions would prevent BIG's use of the code for anything but emergency support, in the event MIMI refuses to or does not support its product. Even if MIMI is willing to do that, BIG may have a problem with those kinds of restrictions. For one, BIG's developers may be doing things that are competitive with MIMI, and because BIG has access to the code without broad rights to use it, MIMI could argue that the competitive products were a result of violations of its source code. Who would prevail is not the issue, as much as the cost and impact of managing such a dispute. This is a problem commonly known in the industry as "contamination". In other words, BIG's developers could be "contaminated" with knowledge of the source code. So, with a potential source code licensing arrangement, we again have a conflict over the restrictions. Of course, money may solve that problem. For example, there may be some price for which MIMI is willing to sell broader rights to BIG. Or BIG could even purchase the MIMI Company. Since MIMI may be willing to sell broader rights for some price, the restriction issue, while conflicting, is not necessarily a mandatory conflict or a deal-breaker.

K&R's MID Chart of Goals		Mandatory (Ends)	Important (Preferred means or ends)	Desirable (Desirable means, some ends)
Conflicting	BIG (Licensee)	"We must have access." (M or E?)	"minimize restrictions" (restricts competitiveness)	
	MIMI (Licensor)	"I can't give you that." (M or E?)	"maximize restrictions" (protect company future)	
Independent	BIG (Licensee)	Protect reputation/ company		
	MIMI (Licensor)	Protect key assets/company		
Joint				

Chapter 10: K&R'S MID™

What might a conversation look like if the parties recognize these conflicts? The discussion would be to explore alternatives to protect against future events for BIG while protecting MIMI's assets today:

M: "Luther, as long as I am supporting you, you really don't need access to the source code. In fact, any license I may give you is going to be so restricted or so costly that you may not want it anyway...right?"

L: "I certainly don't want our developers contaminated...."

M: "And believe me, giving you our source code, our key asset, is going to be expensive anyway."

L: "That's not what I was hoping to hear. But look, I don't really need the code today. In fact, we are coming to you because MIMI is the expert in this type of software. I am concerned that I have access to the code in the event you don't provide us support or can't because MIMI, for whatever reason, is in financial or other difficulty. Or maybe you just cease supporting this entire product line without a follow-on product."

M: "So what are you suggesting, Luther?"

L: "You could give us a license today, but put the source code in escrow with a third-party agent, as is commonly done. Then we would have access to the code only if these events that we are worried about occur."

M: "Hmm. I understand that some of our competitors have done that for their licensees. Of course, we would have to see the terms of such an escrow arrangement, and we would expect you to pay for it."

L: "There is just one other thing with an escrow arrangement. Realistically, if we ever have to access the code, it's likely to be as a reaction to a critical situation, when they are not getting the support they need. It would take my support programmers months to become familiar enough with that code to actually use it and implement full support capabilities. So, I would need a commitment from MIMI and its key developers to provide our programmers with knowledge transfer support if a triggering event occurs; in other words, those key MIMI *developers would provide us with training and interim services in such a critical situation. We could agree to a fee for them in our arrangement."*

M: *"Overall, that sounds like it might work, Luther, but, of course, we have to work out the details...."*

Now the MIMI CEO and Luther Large may be able to negotiate terms of an escrow arrangement that protects MIMI's assets today and allows for BIG's protection in the future. Both companies, using good communication, should be able to find out the problem each is trying to solve, the true goal they are after. Once mutual understanding of those problems develops, it becomes easier to solve what at first appeared to be an irreconcilable mandatory conflict. As a result, the MID would look like this:

K&R's MID Chart of Goals		Mandatory (Ends)	Important (Preferred means or ends)	Desirable (Desirable means, some ends)
Conflicting	BIG (Licensee)	**"We must have access." (M or E?)**	**"minimize restrictions"** (restricts competitiveness)	
	MIMI (Licensor)	**"I can't give you that." (M or E?)**	**"maximize restrictions"** (protect company future)	
Independent	BIG (Licensee)	Protect reputation/ company		
	MIMI (Licensor)	Protect key assets/company		
Joint			- "implementable" escrow source custody arrangement - knowledge transfer	

GOAL TENDER

You're the buyer. You're in a negotiation and the seller says, "I want to do a $4 million deal with you." Visualize K&R's MID™. Is this offer mandatory?

Probably not! The seller *wants* to do a $4 million deal. (After all, who wouldn't?) The deal may be desirable or important. It might be mandatory that the seller move at least $2 million worth of product because of costs, quotas or overhead. But it's probably not mandatory that the seller does the $4 million deal.

Distinguishing between means and goals (ends) is so crucial to a successful negotiation that it deserves additional practice. Try your hand at the following scenario. Answer the questions in the workbook. (Warning: it's a tricky one!)

STOP if you are using the Companion Workbook.

Exercise 10-3: To comply or not to comply

To comply or not to comply

> *ECommerce Solutions Corporation (ECom) is trying to sell additional support systems for HAM's e-services distribution infrastructure. HAM is in the business of disseminating high-quality recipes through a number of printed and electronic mechanisms. The potential business for ECom is $4 million, with about $1 million in software. In the process of doing a proof of concept at the customer site, the technical sales team discovers that HAM has been misusing ECom's database software by deploying over 250 users rather than the 100 users originally licensed.*

> *The cost of 150 extra seats is $500,000, with additional annual maintenance charges of $90,000. The sales manager on the account raises the compliance issue with HAM, to which HAM's procurement*

person responds, "If you do not waive this fee, we may not do business at all this year." A meeting is set up with HAM's line management and procurement staff to discuss the value of the newly proposed solution. After an initially aggressive position is taken by procurement, the conversation with HAM's line manager goes like this:

HAM's line manager: "We would like to be in compliance, but I am not prepared to take a $500,000 hit to my budget."

ECom rep: "I can help you here. ECom has a product that can help you manage your license inventory. It enables you to monitor compliance with your license agreements. In these days of focus on corporate compliance, we're sure that your organization will benefit when audits are done."

HAM's line manager: "I still don't want to take that $500,000 hit."

ECom rep: "I understand what you're saying. There are at least four different alternatives as I see it..."

These were the suggestions:

First and least desirable, HAM can stop using the product for the additional users.

Second, work out a payment plan for those 150 licenses.

Third, ECom can help you obtain financing.

Last, if I can get your commitment for additional business, I may be able to get internal clearance to adjust the fees due.

HAM's line manager: "Such as?"

ECom rep: "We could apply some of the discount we already gave you for the additional order to the money owed, assuming there is a commitment on HAM's part."

HAM's line manager: "Well, I still don't like it, but I appreciate your approach and we will discuss it internally."

Here are the issues we identified. You may have others:

Chapter 10: K&R'S MID™

1. **1.** Payment

2. **2.** Compliance with this contract

3. **3.** Future compliance

4. **4.** "Face"

5. **5.** Continued use of software

6. **6.** General enforcement of licenses

7. **7.** Budget problems

8. **8.** Future $4 million deal

9. **9.** Continued good relationship

Here is how we analyzed each of these issues using the questions we posed. Compare your answers to ours.

1. Payment

Payment is always a conflicting issue, but it is not mandatory for the customer to pay nothing. The customer should be willing to pay something based on value. For the licensor, ECom, some payment is mandatory. But for the buyer, HAM, it is desirable but not mandatory to pay nothing. Their willingness to pay will be predicated on value, so you can see a potential trade-off on this issue if value is identified. Payment is an end for ECom. It is a means to be in compliance and obtain license rights for HAM.

2. 3. & **6.** Compliance (with this contract, and future) and general enforcement of licenses

Compliance itself is not generally a conflict. Unless the customer is a thief, they want to be in compliance with agreements they make. How important compliance is to them will depend on a number of factors, including whether or not they are a public company. Being out of compliance could mean they understated liabilities on their balance sheet. If we assume that HAM is a public company, being in compliance would be an end that is very important or even mandatory.

For ECom, compliance is a means to get paid now and in the future. More broadly, ECom is concerned with enforceability of its licensing agreements. For ECom, all of these requests are mandatory goals. We see that compliance in this instance and in the future under this contract is really a joint goal of different importance for each party. ECom may position enforceability of its agreements in HAM's interest. That's because the loss of revenues for ECom from its inability to enforce license agreements could result in poor product support for HAM.

4. "Face"

"Face" is like budget; It is probably an issue for HAM. Someone at HAM failed to monitor compliance. As a result, ECom may have to deal with a different HAM employee not as closely connected to the issue to get resolution. Either that or ECom needs to provide a solution that helps HAM save face.

5. Continued use of software

Continued use of the software is usually in the interest of both parties. Presumably, HAM is getting value from the usage; otherwise, why would they use the software. ECom would like HAM to continue using the software, provided that ECom gets paid. This would result in maintenance revenues and a happy and probably dependent customer. Current customers that receive value are likely to be future customers.

A compliance scenario should be good for suppliers. HAM is currently using the software, which means they find it valuable. Its value proposition is proven by usage, so the supplier should have leverage. The amount of leverage depends on a number of factors: the business impact of the software, whether there is other software that could do the job, how costly and difficult it is to install other alternatives, retrain people, and so on. In most cases, the fact that HAM is using the software beyond expectation is very important, but dependent on price and value. Discontinuing use of the software is probably the least desirable alternative to solving the compliance issue. And technically, some payment would still be due for non-compliant use before discontinuing!

7. Budget problems

For HAM, this is an independent issue. Although for a seller, a budget issue *feels* like it is the seller's issue. That's why ECom offers a number of different financing alternatives. ECom also proposes to use the compliance and budget issue as a lever to get the $4 million additional deal, a sound use of leverage.

8. Future $4 million deal

Clearly, ECom considers this very important, even mandatory. However, the issue is a "red herring" (misleading). If the value of the $4 million deal is not there independent of the compliance issue, how likely is HAM to make the $4 million commitment? So, using compliance as an incentive to get the $4 million only works if the $4 million deal sells itself. The fact that some of the noncompliance payments can be wrapped into that deal can only help, not hurt, the chances of getting that $4 million deal done. So, the compliance issue is a means to aid the $4 million deal, and the $4 million deal may be a means to solve the non-compliance issue.

9. Continued good relationship

Both parties should have an interest in good relationships that create efficiency and value. Using the MID to sort through the issues means that logic and sound reasoning will be applied. This gives both parties the opportunity to minimize or eliminate acrimony, paving the way to sound long-term relationships.

K&R's MID Chart of Goals		Mandatory (Ends)	Important (Preferred means or ends)	Desirable (Desirable means, some ends)
Conflicting	HAM			- Would like to pay nothing (Free is good)
	ECom	- Payment	- Budget (importance depends) - Face (importance depends)	
Independent	HAM			
	ECom	- Future compliance - Enforceability of licenses (could this be positioned in HAM's interest – as joint?)		
Joint		- Systems must gain HAM's compliance now	- HAM would like to be in compliance; (importance depends on other factors) - Continued use of software (importance depends on level of dependency and cost of switching) - Additional $4M deal (value dependent) - Good working relationship - Financing - Payment plan	

Note: Of course, many of the specific issues in this example go away if the software is implemented in a cloud infrastructure environment.

The decision whether something is a means or an end is a matter of reasonable judgment relative to the deal. For example, while compliance for HAM could be a means for them not to violate SEC rules and to clean up their balance sheet, with respect to this deal, it's a goal and therefore an end.

Look back at the MID Chart. You'll see that very little is in conflict, and there really is no mandatory conflict. As a result, a deal to fix the compliance problem should get done. What we have seen is that using the MID effectively is a 5-step process:

1. Identifying the requests of the two sides and the issues raised

2. Categorizing the issues and requests according to the level of conflict (conflicting, independent, or joint)

3. Prioritizing the issues and requests in the order of importance (mandatory, important, desirable)

4. Determining if the issues or requests that are conflicting are means or ends (goals), particularly if they are in the upper left corner of the MID Chart (mandatory conflict for both sides)

5. Evolving conflicting means to non-conflicting ends and providing less conflicting solutions to address those ends

If there are no mandatory conflicts, the deal should get done.

CATCO, RATCO REVISITED

Remember Catco, Ratco, and Infest City? That's right, in the introduction we asked you:

1. What MUST Catco get out of this deal?

2. From Catco's point of view, what must Ratco NOT get out of this deal?

3. What MUST Ratco get out of this deal?

4. From Ratco's point of view, what must Catco NOT get out of this deal?

These four questions ask you to identify the mandatory goals of the two sides. If the mandatory goals are in conflict, the deal should not get done. Now look at what you wrote. If any items in your answer to question 1 conflict directly with an answer to question 4 or an item in your answer to question 3 conflicts directly with an answer to question 2, this deal doesn't close. That's unless what you stated is either not an end (it's a means) or it's not really mandatory.

For example, Catco may have a goal not to allow access to its underlying Intelligent Eradication System ("IES") technology (its market advantage). Well, if Ratco's goal is to get access to leading edge IES technology, this could be viewed as a mandatory conflict. But what problem is Catco trying to solve by not allowing access to its technology? Catco's goal is to maintain or enhance the market advantage that IES technology provides. Therefore, not giving access to Ratco is a means of protecting that goal.

And why does Ratco want access to IES? Will it do the deal without access? It probably wants access to increase its ability to compete. Can this deal be done without access and increase Ratco's competitiveness? Probably. Again, we are talking about a means, not an end.

Are there different means to improve Ratco's competitiveness and protect Catco's technological lead? Of course! For example, Ratco could get rights to sell Trap out of Newco, without access to the technology itself. Or maybe Catco could license the IES technology to Ratco for a fee that compensates it for negative market impact it believes could occur.

That's what the MID is all about. It helps you analyze conflicting issues to enable you to create negotiated solutions.

Chapter 10: K&R'S MID™

WHAT YOU LEARNED IN THIS CHAPTER

- Arranging goals in order of importance is critical to preventing negotiations from being deadlocked over issues that should not be mandatory.

- People have a great deal of difficulty separating the **means** from the **ends**.

- The **means** are alternate ways of accomplishing goals. The **ends** are the goals. There is usually more than one way to accomplish goals.

- Use K&R's MID to prioritize among mandatory, important and desirable goals.

- *Mandatory requests* are the goals that must be met for the deal to go through. *Important requests* are goals or means that matter to either side, but the success or failure of the deal is not likely to rest on them. *Desirable requests* have the lowest priority, but can be used as bargaining chips to get something more important.

- The following two questions help you separate the means from the ends using our MID: Why? What problem am I trying to solve?

- Not all deals can be made—or should be made. The MID analysis forces you to figure out what problem you are really trying to solve, and whether the deal is viable.

- Patience and listening are two keys to success. Listen to gain understanding, not to argue.

CHAPTER 11: NEGOTIATION STEPS

Have you heard this story?

> *A shabby-looking guy goes into a bar and orders a drink. The bartender says, "No way. I don't think you can pay for it."*
>
> *The guy says, "You're right. I don't have any money, but if I show you something you haven't seen before, will you give me a drink?"*
>
> *The bartender says, "Deal!" and the guy pulls out a hamster.*
>
> *The hamster runs to the piano and starts playing Gershwin songs.*
>
> *The bartender says, "You're right. I've never seen anything like that before. That hamster is truly good on the piano!" The guy downs the drink and asks the bartender for another.*
>
> *"Money or another miracle or no drink," says the bartender. The guy pulls out a frog. He puts the frog on the bar, and the frog starts to sing. He has a marvelous voice and great pitch. A fine singer.*
>
> *Another man in the bar offers $300 for the frog. The guy says, "It's a deal." He takes the three hundred and gives the stranger the frog. The stranger runs out of the bar.*
>
> *The bartender says, "Are you nuts? You sold a singing frog for $300? It must have been worth millions. You must be crazy."*
>
> *"Not so," says the guy. "The hamster is also a ventriloquist'.'*

Unfortunately, negotiators can't get someone else to do the talking for them. In this business, *you* have to communicate yourself. Two of the key steps in any negotiation process are **communication** and **persuasion**. When you master these skills, the negotiation becomes a lot easier.

TALK THE TALK AND WALK THE WALK

In some languages, the English word "persuasion" has a negative connotation, similar to "forcing" someone to agreement. (Remember the discussion of connotation and denotation in Chapter 3.) When we use the word "persuasion" we use it with a positive connotation.

The key elements of effective persuasive communication are:

mere communication	←——————→	communication that is <u>understood</u>
credible communication	←——————→	communication that is <u>believed</u>
persuasive communication	←——————→	communication that is <u>valued</u>

These three attributes of communication build on each other—credible communication must be understood and persuasive communication must also be believed. Effective persuasive communication means completing your value argument. Delivering value to the other side is one of the best ways to get them to agree with you. If they perceive that you offer something of value that other alternatives do not, they are more likely to agree with you and get the deal done. Yet, over years of negotiating deals and teaching people to negotiate in a business environment, we have found that people have trouble both articulating and quantifying value to the other side.

Even the first element, being understood, occasionally presents a challenge. This is especially true in a business environment. If we present a solution to the customer, it is our responsibility to ensure that our proposals are clear, and speak to the other side in their context.

Most people should not have much trouble being credible. Maybe that's because we can be credible by pointing to what we know about ourselves and our company. For example, we can easily present a business case showing why the price we are charging is competitive in our industry and consistent with a reasonable cost structure. However, to show the other side why they should pay that price—that our solution will make them more competitive—we must know something about them and about their business. That is more difficult, because we are not as knowledgeable about their business as we are about our own. When you know their business, their terminology, you can frame your argument in terms and values that they understand. You must get the information you need about them—skills we discussed in Chapters 8 and 9.

Let's look at this graphically.

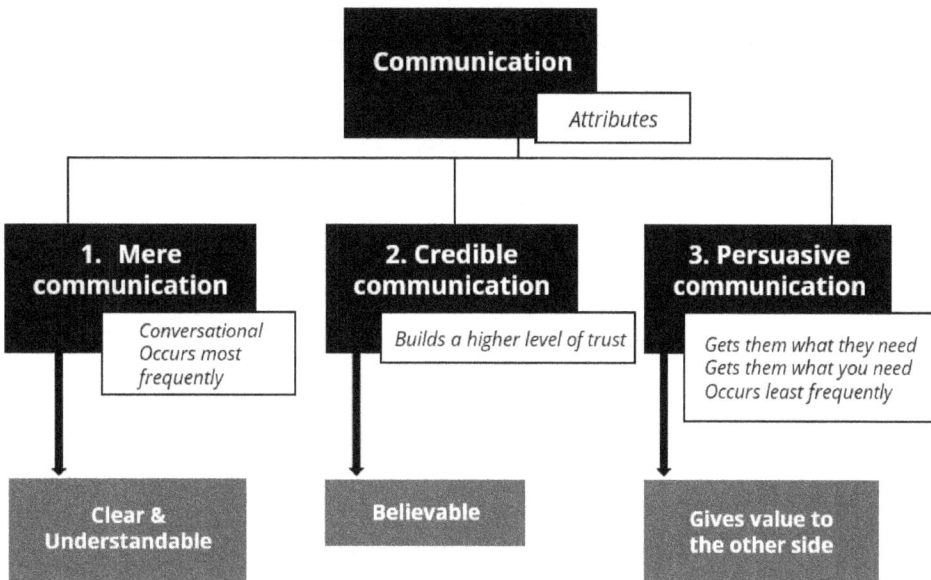

Figure 1: The Three Attributes of Persuasive Communication

Chapter 11: Negotiation Steps

Persuasive communication requires value.

Complete the activity in the workbook to analyze what effective, persuasive communication can achieve for you.

STOP if you are using the Companion Workbook.

Exercise 11-1: If at first you don't succeed...

Effective persuasive communication recognizes two kinds of values:

- Company value to the other side

- Personal value to the representative or decision maker of the other side

You generate company value by making the deal beneficial to the customer's organization. This helps their company's bottom line or revenues, improves their return on assets, expands market share, reduces costs, etc. In other words, their company is better off for doing the deal. Their corporate or organizational goals are achieved.

You generate personal value by providing legitimate individual benefits to the people on the other side who are involved in the transaction. (As mentioned, this does not include illicit personal benefits or commercial bribes!) This value could mean increased sales for the sales executives, funding for development executives, or other results from doing the deal with you that helps them achieve their goals.

It also includes personal elements such as providing your counterpart with credibility and face with their peers, bosses, and others. Providing personal value also raises your Negotiation Capital for the next deal you make.

More often than not, personal value is simply making the job of your counterpart easier.

Here's a familiar example, but from a different perspective:

> *A few years ago, Mladen had a client that we'll call "Shore", whose small sales team was getting frustrated with a customer, "Land", because they couldn't close a deal. Land's director, Sally, managed a line of business that would benefit from Shore's solution. She was an advocate for Shore based on her understanding of the benefits that were communicated by the Shore team.*
>
> *As Mladen puts it: "When discussing this deal with Shore's team, I asked what the roadblock might be. The team answered that Sally needed the CFO's approval of a business case to justify any expenditure in excess of $500,000. This deal as proposed was for $1.8 million. When asked if they had seen Sally's business case, the team said no. I suggested that the team build the business case for Sally. They did, and the quantification of value showed a great ROI for Land. When the team presented the business case to Sally, the first thing she said was, "Can I have a copy? It's what I haven't had time to focus on."*
>
> *She took the work the Shore team had done and got the CFO's approval for the deal.*

The K&R Deal Forensic

Once the Shore team understood why the deal was not closing, they used several good techniques to address the issue.

1. They quantified value for the customer.
2. The Shore team provided personal value to Sally by making her job easier.
3. Sally's productivity and stature inside her company were enhanced, as was the overall relationship between the companies.

The bonus of their approach is that quantified value enabled a successful internal negotiation for Sally. More on that later in this book.

K&R's Negotiation Dynamics

- Hard work gains you knowledge.

- Knowledge gains you credibility.

- Credibility gains you leverage.

Leverage used wisely leads to desired results and successful relationships.

Axiom: Information and preparation are key to winning negotiations.

The dynamics are observations that overlap with what we show in the K&R Leverage Cycle. A successful deal has the attributes of these dynamics. And the bridge between knowledge and credibility is the confidence you gain from knowledge. Remember in the Dataco example in Chapter 8 how the person with knowledge was perceived versus the one without knowledge? When you have knowledge, confidence, and credibility, you have the assets with which to make rational arguments that persuade the other side to voluntarily move to your way of thinking (positive leverage). Rather than having to rely on intimidation, you have the makings of good long-term relationships based on value the other side perceives.

BARGAINING AND PERSUASION TECHNIQUES

The context in which your value arguments get communicated will dictate your bargaining success. You use certain persuasive techniques to deliver value arguments. These make up your planned concession and negotiating strategy. Here are Eight Persuasion Techniques we consider primary:

1. Principled concessions

2. Logical reasoning and use of resources

3. Stimulus/response

4. Repetition (Persistence)

5. Conscious use of "you"

6. Positive inducements

7. Negative inducements

8. Tactics

Let's look at each one more closely.

1. Principled concessions

From a persuasion standpoint, principled concessions are closely related to logical reasoning. A concession given without rationale has no logical underpinnings. The unprincipled concession raises questions on the part of the other side. For example, they might think, "Why did they make that concession?" or "Are we missing something? If they gave in so easily, there must be more to give."

On the other hand, a principled concession related to value tends to reduce cycle time and get closure more quickly. For example, if you make a credible offer with quantified value that contains five elements, and the other side doesn't accept the second element, you may reduce the price relative to the value of that element providing you agree that the element has less value then you originally assumed. The other side doesn't have to ask why you made that concession or question whether it was too easy. You explained it. In return for that principled concession, you would expect to receive a principled concession: agreement from the other side based on the remaining value of the transaction. What could be more fair and respectful?

2. Logical reasoning and logical use of resources

Is logical reasoning an art or a science? We explored this issue in Chapter 2, "The Art and Science of Negotiation". We concluded that logical reasoning contains elements of art in its creativity, science in its precision and predictability based on experience. In Chapter 3, we spoke about errors in logical reasoning that generate illogical conclusions. Logical reasoning, in contrast, generates logical conclusions.

Most effective persuasive communication has elements of logical reasoning. You take generally accepted assumptions and draw logical conclusions based on explainable rationales. That's what building a business case is all about. This is the *science* of negotiation.

Here's the art element: Your team develops a great financial business case, principally done by the finance MBA on your team. You also have a compelling technical analysis for your solution, done by your technical guru or development engineer. It's time to take the financial case and the technical presentations to the other side. The *art* is how we deliver the logical reasoning. Should you make the presentations? Do you have the financial and technical expertise? Can their technical and financial experts trip you up? An excellent alternative is to use the experts on your team. Rehearse and challenge them as part of the preparation. As long as you trust the people who developed the plan to stick with it and to present it clearly, then they should make the presentation.

The following story from our files illustrates the melding of art and science into the essentials of logical reasoning and principled concessions:

> *Brian Johnson, a mid-level manager, was not what you would call a wise negotiator. The man just didn't get it. But judge for yourself.*

Chapter 11: Negotiation Steps

We were all set to close the deal when Johnson decides to give the customer 10% more discount than the customer had requested.

This translated to $258,000. There was no reason on earth to make this concession: the deal was all but signed. Nonetheless, Johnson sends us a note saying, "Give them 10% more." We decided to ignore the note.

A team member calls and says, "Did you see Johnson's note asking that you give the other side 10% more?"

"No," we reply.

Another teammate calls and says, "Did you see Johnson's note asking that you give the other side 10% more?"

We say, "No."

The lawyer on our team calls and says, "Did you see Johnson's note asking that you give the other side 10% more?"

We say, "No."

Only the lawyer realized that by denying that we had seen the note, we were guilty of ignorance rather than insubordination. We went ahead and finalized the deal at the higher rate. Ignoring the note to make an unwarranted concession not only saved the company money. It probably saved time, too, since making an unprincipled concession would have raised questions about credibility and could have triggered requests for additional concessions.

Sometimes you take a risk. In this situation, we decided the risk was worth it. That's not to say that we recommend you do the same thing. But good judgment is critical, and in this case, good judgment dictated avoiding an unprincipled concession.

The K&R Deal Forensic

What made Harvey's actions work was a combination of art and science:

1. It was predictable that giving the other side an additional 2% would be viewed as an unprincipled concession.
2. An unprincipled concession would probably result in prolonging the negotiation.
3. We were more likely to close the deal sooner by sticking to the existing terms.
4. The creative handling of the middle manager's request was art.
5. Ultimately it was predictable (science) that the middle manager would not complain after the deal was closed for a higher price. But remember, we don't endorse insubordination (unless you are independently wealthy or 100% sure you are right.)

3. Stimulus/response

With this bargaining technique, you provoke the other side to get a response. Your goal is to persuade the other side to make a concession. The following anecdote from our files is a case in point.

> *The customer forced us to offer a price sooner than we wanted to, since we had not discussed all material terms. Harvey said, "You are forcing me to give you a price now before all terms have been finalized. Therefore, this price is based on the contract terms, which are currently in front of you. Since price is directly related to terms, if any terms change, the price can go up or down. For example, the current warranty is twelve months. If you want that changed to 36 months, the price will go up. Conversely, if you want a three-month warranty, the price will go down. Do we agree?" The customer agreed. We gave them a price.*

One week later (on December 20th*) the customer came back with, "I'm still not happy with these terms. And don't tell me that if we make changes the prices are going to go up."*

Now, our team would receive a closing bonus if we could get the deal done by year's end. Some of us also wanted to take a year-end vacation. Our fear was that if we gave in on these issues without relating back to price, the customer would reopen many other important terms.

Making a decision on the spot, Harvey stood up and a little more loudly stated, "May I remind everyone of the deal we agreed to when we gave an early price. If terms change, the price can go up or down. Does anyone remember that or are we having convenient memory loss?" He looked upset, but in control. His "controlled indignation" got the customer's attention because during months of negotiation they had never seen him upset. Harvey took the high moral ground because we had a deal. He walked out of the room.

Outside the room, he made a call to check his voice messages.

Their lead negotiator came out to the hall a few minutes later to calm him down. After a few minutes of talking, the lead negotiator agreed that since price could change with the terms, they would leave the terms as they were and sign the deal.

The team members got their signing bonus and everyone got their vacation.

What if the stimulus/response had failed to get the desired result? Since we had a compelling value proposition, the deal would have closed anyway; however, it may have taken longer since the value of the additional terms would have to be negotiated. That would have put additional pressure on our team to make concessions to preserve everyone's year-end bonuses.

The K&R Deal Forensic

How did this tactic work?

1. First, we understood the risks of using the tactic and determined they were minimal.
2. When you are right on solid ground, controlled indignation can be an effective stimulus/response tactic.
3. There had been clear communication on the issue of the relationship between terms and price.
4. The rationale behind Harvey's controlled indignation was based on that agreement and was, therefore, on solid ground.

Creating a stimulus to gain agreement is a positive method of persuasion when it is rooted in principle. And it can work if the value proposition is there. However, many negotiators use negative stimuli as tactics in negotiations. More on that later.

4. Repetition (Persistence)

Below are a few really important things that our kids have taught us:

1. Ask why until you understand.

2. Just keep asking for the cookie until someone gives you one.

3. Don't give up. Persistence pays.

As you learned in Chapter 6, Principle #5: "Negotiation is a continuous process." Children are natural negotiators because they are persistent and focused. They repeat their arguments, which are often very simple: "I want...I want...I want." In business, our arguments are more complex, but the same techniques of repetition and persistence pay off.

Mladen used persistence when negotiating with a contractor on a personal matter, as the following anecdote shows:

> *I was in discussions with a contractor over the demolition and reconstruction of my deck. It was fall, so I wanted to get it done before wintertime. The contractor I wanted to use was very busy and suggested that we wait until spring. But I am one of the few people who use the grill all winter and did not want to risk using the deck with snow on it, if it was not redone by mid-November.*
>
> *We had already agreed on the materials, design and price, so the negotiation was all about schedule. The contractor had one advantage – he knew that I wanted to use him because of previous positive experience. I explained why I needed the deck done now. He apologized and said that he did not want to promise he would get it done, when he knew his calendar was too tight.*
>
> *I waited a day and called him back, saying, "Jay, I respect the fact that you don't want to promise me something you can't live up to. But you trained me to appreciate your services, and I think you are the best around. I really don't want to go to look for someone else." He said he would think about it.*
>
> *A day later I ran into him in town and asked again, restating how important this is to me. I suggested that I would speak with one of his other clients whom I knew, to see if the schedule could be reprioritized. He said he would think about it.*
>
> *He called me the next day, and I restated my offer to call his other client. He said that would be OK. I called the other client who agreed to postpone his schedule until my deck was done.*

Persistence is necessary not only to ensure communication is clearly understood, but to communicate that something is truly important. Something communicated infrequently naturally gets lost over time, losing importance in the eyes of the other side.

Chapter 11: Negotiation Steps

> **Patience and persistence are key to persuasion in both our business and personal lives.**

5. Conscious use of "you"

"You" is a loaded word. Think about this: Sam says to Jane, "You said...." How is Jane likely to react regardless of what follows the word "said"? Most people would say, "defensively". If Sam is disagreeing with a position Jane's company is taking, how does use of "you" help him? By using "you", he's identifying Jane with a position with which he disagrees. He is forcing her to defend that position! Wouldn't Sam be better off saying something like, "XYZ (Jane's) company has taken the following position... Please understand that we feel differently for the following reasons..." Now Sam is giving Jane the opportunity to have an independent opinion and evolve to his position based on rationale. He is not forcing her into a knee-jerk defense of a position with which he disagrees. The famous treatise from Harvard's Fischer and Ury, *Getting to Yes*, refers to this concept as separating the people from the problem or issue.

> **When trying to persuade, use "you" consciously to confirm a position you agree with. Avoid using "you" when you want to change someone's mind.**

6. Positive inducements

As you learned earlier, predictions and promises are forms of positive inducements. To some extent every business case is a prediction of what we believe will happen in the future. But when you make promises in a negotiation, the other side will ask to have it in writing. Always focus on K&R Principle #4: "Concessions easily given appear of little value." If the inducement is in the form of a concession, it should have value and should be given for a good reason (principle) to be an effective motivator in a negotiation. Make sure the concession is related to identified problems in the negotiation. Otherwise, you will not get the other side to come closer to agreement. You will not be motivating the behavior your inducement (the concession) is trying to get. All you will have done is erode your position and receive nothing in return.

7. Negative inducements

Using a negative inducement is a conscious decision to get the other side to worry. This is the "stick" in a negotiation: warnings and threats. A warning is a prediction (positive inducement) stated negatively. For example, "If we don't do this deal today, here's what you stand to lose over the next three months". There are many studies showing that people act with more urgency in relation to risk than reward. Threats are associated with risk but could be detrimental to relationships. Therefore, you should consciously decide whether to use a positive or negative inducement. But never make a threat you are not willing to carry out. Remember what you learned in Chapter 10 about using K&R's MID™. Credible threats are centered on mandatory goals, the deal breakers.

8. Tactics

Tactics play such an important role in the negotiation process that we devote an entire chapter to them (Chapter 13). Below is a brief example to whet your appetite. We call this classic Mladen story "If the Sun Don't Shine".

> *I was in a negotiation with a well-known tech company we'll call "Bright". The meeting was in the afternoon, in their offices in California. When we arrived, they took us to the conference room for the meeting, noting how difficult it had been to get a conference room on this busy week in their offices. The room was on the west side of the building with windows facing the afternoon sun. It was hot and bright. The three members of the Bright team were already seated at the small oval*

table, facing the door. This meant that we would need to sit looking straight into the afternoon sun.

I noted the brightness of the sun as we sat down, to which their lead negotiator smiled and remarked "Isn't it a great sight?" We put our bags in the corner, and our materials on the table and, as we sat down, I asked "could we have a moment please; I would like to discuss something with my team for only five minutes".

As soon as they exited the room, I asked our group to switch positions with the Bright team. We were now seated facing the door.

As they reentered the room, they immediately noticed that we were seated facing the door, and they would need to sit facing the sun. Their lead negotiator (to his credit) and the rest of his team laughed, as I said, "We are not used to the glare of the sun, and thought you would enjoy the sight". As we began the meeting, my Bright counterpart asked one of his people to find an alternate room. Within the hour, we were all more comfortable in a room on the east side of the building.

At the end of the day, their leader turned to me and said, "Thank you for a very productive day". My tactic had worked: they realized that we would be respectful, business like and firm while getting the job done. We built business respect for a relationship that has lasted to this day.

DO THE LEG WORK

Remember this K&R negotiation dynamic:

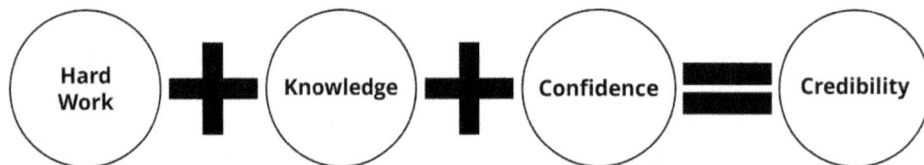

Figure 2: The Credibility Equation

Chapter 11: Negotiation Steps

People are products of their experience. That's why in our information gathering, it's important to find out about the past experiences of the decision makers and negotiators on the other side. It's also important to understand the experiences and motivations of the members of your own team. Some of the questions you should explore are:

1. What experience do these people have with the negotiation process?
2. Have I ever worked with them? Have people I know and trust worked with them?
3. What education do they have?
4. How would I rate their level of expertise? Excellent, good, fair, poor?
5. What is their negotiation style? For example, are they adversarial or problem solvers? Or are they confrontational and difficult or laid back and relaxed?
6. Are they decisive or not?
7. What deals have these people negotiated in the past?
8. Were these deals a success or a failure? Why?
9. What do the participants have to gain?
10. What do the participants have to lose?

Gathering this information helps foster relationships by satisfying concerns that are based on personal experiences. That, in turn, makes it easier to solve difficult negotiating issues.

Chapter 11: Negotiation Steps

Study the following chart:

Figure 3: Preparation is Key

PLANNING, UNDERSTANDING, SETTING

The negotiation preparation process can be summarized into three overlapping stages:

1. Planning

2. Understanding Business Goals

3. Analyzing the Negotiation Setting and Environment

The *Planning* stage includes:

- Information gathering: M.O.R.E.

- Information management

- Information utilization: shifting the leverage

Understanding Business Goals includes:

- Assessing the measurable business impact for each side of delivering the sought business outcomes (value-based leverage)

- Analyzing the credible offer: understanding the NSR

- Using the MID analysis—means versus ends and developing a principled concession strategy

Chapter 11: Negotiation Steps

Analyzing the Negotiation Setting and Environment includes:

- Using natural and acquired skills

- Tapping experience, culture, and other factors

- Determining what team resources should apply to both sides

We've already mentioned how the impression you convey can say as much—or even more—about you as your words. Exhibit confidence about what you have to say and your ability to say it well. Don't leave the door open for the other side to doubt you; for instance, if you cite a key fact, indicate the source or how you verified it.

Choose your words carefully. For example, saying, "I am sorry to have to tell you this..." probably tells the other side that what you are about to say is either bad or something you will not use voluntarily for your advantage. If it is bad, let them decide, and if it is good for you, why create an expectation that you won't use it? You probably shouldn't begin your negotiation with, "I'm not used to doing this, so you'll have to give me a little latitude," or "Unaccustomed as I am to negotiating a big deal like this one..." These approaches expose your weaknesses and potentially your credibility. You may be looking for sympathy, but there is a really good chance the other side will perceive an increase in their leverage and become less flexible. That is not to say that the humble approach can't be effective. In fact, asking politely and humbly when you don't know something can be a very powerful way of getting the other side to give you information.

CULTURE CLUB

- Frank Perdue's chicken slogan, "It takes a strong man to make a tender chicken" was once translated into Spanish as, "It takes an aroused man to make a chicken affectionate".

- Scandinavian vacuum manufacturer Electrolux used the following slogan in an American campaign: "Nothing sucks like an Electrolux."

- When Parker Pen marketed a ballpoint pen in Mexico, its ads were supposed to have read, "It won't leak in your pocket and embarrass you." But the company translated the wrong word for embarrass, "embarazar" (*colloq.*: to impregnate), so the ad read: "It won't leak in your pocket and make you pregnant."

To succeed as a negotiator, you need an understanding of other cultures, beliefs, and patterns of acceptable behavior. Communicating with people from different cultures is crucial if you want to make good deals. Your knowledge must extend far beyond just avoiding language bloopers!

In the past, America used the metaphor of the "melting pot" to describe the assimilation of different cultures into the whole. Today, the metaphor has become a "mosaic" or "quilt" to more accurately reflect cultural diversity.

Are you attuned to successful intercultural communication? Prior to negotiating any international deal, take the test in the workbook to find out.

STOP if you are using the Companion Workbook.

Exercise 11-2: We are family

Even if you never leave your own country such as the United States or even your own hometown, you'll be negotiating with people whose culture and background differs from yours. That's because there's cross cultural movement between peoples and economies across the globe.

In the last American census, 36% of people chose to identify themselves as minorities. There are almost 3 million people in Germany who have at least one parent who emigrated from Turkey. This makes them the country's largest ethnic minority. There are 1.3 million Indian South Africans. Durban is the largest "Indian" city outside of India itself! The People's Republic of China (PRC) officially recognizes 56 ethnic minority groups in addition to their Han majority. As of 2010, the combined population of officially recognized minority groups comprised 8.49% of the population of mainland China. Get the point?

Cultural awareness has a tremendous impact on negotiations. Successful negotiators carefully and thoroughly prepare to meet with teams from other cultures. The following suggestions can help you communicate more successfully with people from other cultures.

1. As a negotiator, you must always be cognizant of geography.

This factor is so crucial that we devote an entire workshop to cultural differences that will drive changes in negotiation approaches. Never forget where you are and the effect you are creating with your cultural norms.

2. Understand the etiquette and negotiation approaches of other cultures.

For example, negotiations in Tokyo take longer than negotiations in New York, since the expectation is to work on the relationship first. If you understand this basic difference in negotiation culture, you can set the right expectations with your management. That, in turn, puts less pressure on you to show instant results, and allows you to take the appropriate time to form a good relationship.

3. Learn key nonverbal clues.

Learning another culture involves more than words—you must learn body language as well, as we spoke about earlier. For example, making eye contact is considered a sign of respect and integrity in some cultures, such as the U.S., while in certain places in the Middle East or in Korea, it could be viewed as threatening or a sign of disrespect.

    4. Use the other country's language carefully, if at all.

To foster good relationships, it is important to learn a few phrases in local language. Most people around the world appreciate the attempt. But be careful.

President John F. Kennedy got tripped up with his famous "Ich bin ein Berliner" line. JFK *wanted* to say, "I am a Berliner," but he is often credited with saying, "I am a jelly donut." That's because the correct phrase is *Ich bin Berliner*— "I am Berliner" without the article "a". Adding the article results in the slang phrase for *a donut.* While this story is widely disputed, the point is a good one: Beware of foreign slang.

If your company has an office in another country, arrive early so you can meet with the local office members and learn more about local mores and behaviors. If necessary, take your own translator when you travel to another country on business. Discuss with the translator the nature of your work abroad, the negotiations you will be having, and any technical terms you will be including. A good translator can also help you interpret nonverbal behavior and foreign cultural issues.

    5. Use the country's measurement terms.

The U.S. is one of the few countries in the world that does not routinely use metric measurements. If you are negotiating in a country that uses metric terms, translate your figures into metric. Then everyone is on the same page when it comes to size. Also, in dating documents, most of the world uses the DAY/MONTH/YEAR format, not the month/day/year format used in the U.S.

    6. Check your timing.

Chapter 11: Negotiation Steps

If you are negotiating in another country, try to adjust the meeting time to accommodate everyone's internal clock. Try to avoid negotiating when you or members of your team are jet-lagged.

7. Reinforce the need for intercultural communication.

Address cultural diversity head-on to reassure customers you're making a sincere attempt to understand their culture. You can cite statistics, examples, and anecdotes that stress the need for international understanding. Do a simple internet search.

8. Be aware of differences between British and American English.

All English is not the same! British and American English use different terms to mean the same thing. People in Europe, Asia, Africa, and the Caribbean have learned British English, not American English. Study this chart to avoid embarrassment:

American English	British English
Legal holiday	Bank holiday
White-collar job	Black coat job
Attorney (in non-court work)	Solicitor
Attorney (who goes to court)	Barrister
Ground floor	First floor
Busy signal	Engaged tone
Sell (quickly)	Flog
Elevator	Lift
Mail	Post
Call (by phone)	Ring up

American English	British English
Car hood	Bonnet
Subway	Underground
Underpass	Subway
Truck	Lorry
Car trunk	Boot

9. Give yourself the edge.

Think of this: People are flattered by your knowledge of what matters to them. Your preparation should include learning about their current events and history. While you are still home, you can get a lot of relevant historical and current commercial information from the trade missions and embassies of the countries you want to visit, or missions' websites. When you are there, your embassy can provide you with additional information. While you are on the way, read local papers (if translated), the *Asian Wall Street Journal, The Economist,* and other international versions of worldwide publications for relevant local data.

REALITY BITES

As we draw the negotiation preparation summary to a close, keep these ideas in mind:

1. Leverage shifts, as we explained in Chapters 4 and 5. The buyer usually has the leverage in the beginning of a negotiation. That's because they have alternatives and the money the seller wants. As the buyer gets more dependent on the seller's technology or other inducements, the leverage shifts. That's one of the reasons why smart buyers try to lock in as many seller-guarantees as they can the first time around. Smart sellers also try to maintain as much flexibility as possible. That's what a good negotiation is all about.

2. Almost all actions in negotiation impact credibility, leverage, or both.

3. Usually the key issues for buyers and sellers are the same, but with a different view. For example, when the buyer feels they will become dependent on the seller, they want guarantees from the seller such as delivery, supply, right to follow-on products, and fixed prices. Since the seller knows that buyer dependency usually results in a long-term relationship, the seller wants flexibility over issues such as service resources or time to address supply conditions and component prices that may be unpredictable over a long period.

Chapter 11: Negotiation Steps

WHAT YOU LEARNED IN THIS CHAPTER

- Two of the key components in any negotiation process are **communication** and **persuasion**.

- Effective persuasive communication gives company value and personal value.

- Bargain and make principled concessions by using logical reasoning, stimulus/response, repetition, positive inducements, negative inducements, and tactics.

- Hard work (patience and listening) gains you information and knowledge which results in confidence and credibility.

- The negotiation preparation process can be summarized in three steps: planning, understanding business goals, and analyzing the negotiation setting and environment.

- To succeed as a negotiator, you need an understanding of other cultures, beliefs, and patterns of acceptable behavior.

- Leverage shifts. Almost all actions in a negotiation impact credibility, leverage, or both.

- Key issues for buyers and sellers are often the same, with a different view.

CHAPTER 12: NEGOTIATOR'S RESPONSIBILITIES

A member of my team tells the following story. As you read it, decide what it reveals about the negotiator's responsibilities.

> *I was negotiating with a company we'll call "Banana".*
>
> *Negotiations with Banana's team were creeping along at a snail's pace. Banana's executives thought my team and I were deliberately making the negotiations go slowly. In truth, we couldn't make any progress because Banana's lead negotiator and individual team members were arrogant, insulting, and making absurd demands. They stonewalled the process. As lead negotiator, I was attempting to move the negotiations forward at an appropriate and timely pace.*
>
> *Banana's vice president walked into the room for the first time as their lead negotiator and I were discussing cancellation charges. Banana's lead negotiator sneered at me and my team, and said,*
>
> *"Why do you have order cancellation provisions? Aren't you going to build all the units we're going to buy and put them in a warehouse? Then we'll have available supply, and order cancellation charges will not be relevant."*
>
> *Now, we all know that executives lose jobs over excess inventory. In a low voice, I turned to their vice-president and said, "Would you build and stock millions of dollars' worth of inventory for 14 months?"*
>
> *In a situation like this, it's best not to ask a question if you don't know the answer. Of course, I knew the answer to this question. The vice-president predictably replied, "No, of course not."*
>
> *I then said in a low voice, "Now you understand why this negotiation is taking so long." I had created credibility. The V.P. understood that the*

negative atmosphere was being created by his own people. It was his team that was slowing the process. Banana's vice president replaced his lead negotiator. The deal was closed shortly thereafter.

The K&R Deal Forensic

What enabled this deal to close despite some stonewalling by the other side?
1. Our teammate understood his responsibility to get the job done as efficiently as possible.
2. He maintained his credibility in the negotiation.
3. Rather than getting caught up in the emotional battle, he used good logic and business rationale to achieve his goals.

LEADER OF THE PACK

The K&R Negotiation Method™ divides the negotiator's responsibilities this way:

1. Understand and communicate negotiation goals.

2. Establish the negotiation process with the K&R Negotiation Method.

3. Determine the resources (types of skills) that are required.

4. Establish support roles and rules.

5. Earn trust with the team—including management.

6. Address and manage internal conflict to bring in the best overall deal.

This process works for a large company, a small company, or even a household looking to make a major consumer purchase such as a car or new house. Let's look at each responsibility in detail.

UNDERSTAND AND COMMUNICATE NEGOTIATION GOALS

As lead negotiator, you have to understand and then fully communicate management goals to the entire team. Think of yourself as the conduit, smoothly passing information along the line. Information that you must communicate to your team falls into these broad categories:

Parameters	Guidelines regarding price ranges and the flexibility of terms
Expectations	What you are trying to accomplish (what are management goals)
Deadlines	The expected or mandatory time frames to get the deal done
Process	How the negotiation will be conducted

This leadership approach instills trust and loyalty within the team, driving them to common, desired results.

ESTABLISH THE NEGOTIATION PROCESS WITH THE K&R NEGOTIATION METHOD™

"Process" refers to how the negotiation is going to take place. When it comes to process, you are concerned with when, where, who and how. As you will see in a later chapter, the goal of the macro agenda is to map the process from beginning to closure and beyond, with each step in the process moving you closer to getting the deal done.

Here are some questions to consider at the early stages of the process:

1. How is the team selected? Who gets to choose the members of the team?

2. How many people will be involved in the negotiation, directly or indirectly? (In most negotiations, there are only a few people directly involved, but more may be indirectly involved by interacting on the technical side, providing data or discrete areas of expertise.)

3. What is the role of each person?

4. What will be the chain of command?

5. Who is empowered to make negotiation decisions? For example, should everyone involved make decisions? Only subject matter experts? Or only you as the lead negotiator?

6. Is there one unified negotiation position?

7. If not, what is your process for developing a unified negotiation position?

8. Who will be your second seat, if anyone? Is there anyone else who should attend every negotiation meeting or call, and who could step in as lead, if necessary?

9. When will negotiations begin? When will they end?

10. Can you agree on start and end dates? Internally? With the other side?

11. How long should the negotiation take?

12. Will the negotiation take place via telephone, teleconference, or in

person?

13. If you are meeting in person, will you meet in your location, their location, a neutral location, or some combination of the three?

14. Are you prepared for the iterative agreement draft process?

All of these questions need to be addressed early and verified or evolved at various subsequent stages. Of course, the depth of consideration of these elements depends greatly on the size of and importance of the deal being negotiated.

Why is controlling the drafting of documents so important? Controlling the draft helps you manage the agenda by controlling timing of changes and the flow of issues. You want to use your contract because it has already been approved by the various functions of your company. This saves valuable time on your end. Even more importantly, it accurately reflects what you can and cannot do in the relationship, which should increase your leverage. However, this leverage is only sustained by providing good rationales for the positions reflected in the document. Other advantages are:

- Since they are your terms, you can explain why they are in the contract. Therefore, you will increase your credibility

- When momentum to closure occurs, the remaining terms represent your position. This is discussed later in this chapter.

ESTABLISH SUPPORT ROLES AND RULES

To succeed, you must engage the right resources. Assemble a team that suits the needs of the situation. You may need strong people from sales, services, finance, technology, and legal. Above all, the lead negotiator has the responsibility to ensure that all members of the team support a unified position for the company. This does not mean that you want all "yes" people. Rather, you want independent thinkers who bring new perspectives and add value to the team's planning process. You want lots of debate internally to achieve a unified position externally. You want dealmakers, not deal breakers.

It's not enough just to have these people on your team—they have to be available when you anticipate needing their personal attention or attendance in a negotiation. You don't always know when an issue will come up that requires the expertise of your financial person, tech person, lawyer, or some other subject matter expert. Regular communication with the key members of your team is critical. When you have the appropriate resources, and they are prepared, it allows you to make decisions to move the negotiation process forward. Based on the goals and subject of the negotiation, the lead negotiator has to determine which functional skills will be required, and may then have to go inside the organization to recruit people with those skills. An internal kickoff meeting should take place before you meet with the other side. This meeting can be brief (an hour conference call) or lengthy (days long) depending on the complexity of the transaction and negotiation issues, the number of participants, and other preparation issues we have discussed throughout.

Take a moment to reinforce what you have just read by creating your list of issues in the workbook.

STOP if you are using the Companion Workbook.

Exercise 12-1: Planning pays off

Here's our non-exhaustive list. At this internal meeting, you as the lead negotiator should address the following issues:

Chapter 12 Negotiator's Responsibilities

- Your goals for the deal, including management's view

- The goals of the other side

- The varying degrees to which you anticipate satisfying their goals

- The proposed timeline for the deal

- How the process will work

- How you and the team will communicate

- How you will communicate with management

- Who will actually write and revise the contract as you negotiate

- Who will directly participate during the initial rounds of negotiations, whether in person or remotely

- What will be each person's expected role in the negotiation

- What your NSR will look like

- What issues both sides are likely to consider most important

- The various ways these issues can be resolved

- What will be the preliminary order of priorities on an issue (MID™)

- What your leverage points are

- What their leverage points are

- How resources will be used

- What concerns members of your own team are likely to have

- What it takes to get these concerns resolved internally before they impact the external negotiation

- The unified position(s) that the entire team will support at each stage

Again, the level of detail on these issues, as well as the number of people involved, will vary greatly depending on the type of the deal, its size, and its importance. We need to make conscious trade-offs in the process. Getting support from team members and management in a strategy requires similar reasoning and persuasion as with a customer. You need to explain how your strategy is the best alternative to accomplishing their goals.

But what happens if, in the middle of the negotiation, your executive gets a call from the other side claiming that you are not being reasonable on a number of issues? Will your executive confer with you or try to solve the problem without consulting you?

Mladen tells this story:

> *I was in a negotiation with a Japanese company we'll call "Nagoya" whose senior executives wanted to meet with my client's senior management. Luckily, I had been keeping my client's management well informed of the status of the negotiation. During the meeting, Nagoya's Executive Vice President mentioned a few matters they wanted to have addressed. My client's E.V.P. and general counsel politely said, "We would be happy to discuss any matter with you, but on this deal, I am not as knowledgeable in all the details as Mladen. Therefore, we will rely on him to resolve your concerns!"*

That statement made the rest of the negotiation much easier. Why was Mladen's client willing to give Mladen such strong support? Why was it in his best interest to handle it the way he did?

1. If he had addressed these issues, he would have become the lead negotiator for all difficult issues. He was too busy for that.

2. Mladen had earned his trust by communicating with him throughout the negotiation, so he was less likely to get blindsided with requests from Nagoya. This gave him confidence that things were being handled appropriately.

EARN TRUST WITH THE TEAM AND MANAGEMENT

To earn trust, you must maintain credibility. This is not as easy as it sounds. Rank has privilege—how do you work with management, since they outrank you? Here's how: You earn your credibility through action and communication; then do your best to persuade them what path to take. After that, you get direction and follow what they tell you—unless you're independently wealthy. Then you can follow your own star!

The following story from K&R's files illustrates the two-way nature of dealing with management when they don't understand the importance of communicating credibility and maintaining trust with the team.

> *The Shaky Group was on the verge of bankruptcy because their cost structure was out of control. They wanted product for resale from our client which we'll call "Stable." Shaky wanted a really deep discount to help solve their cost problems.*
>
> *Stable responded, "If we gave it to you for free, it still wouldn't save you because your cost structure is so bad. It is unreasonable to ask us, as a trusted supplier, to solve your infrastructure cost problems by selling below market at a loss." Shaky also requested that the requirement for them to differentiate their solution from the products sold by Stable be removed. Stable's goal of selling its products to Shaky for resale was to expand its market. Therefore, the differentiation requirement was critical to Stable, which sold the same products to end-users.*
>
> *After considerable negotiations, the parties were ready to close. Stable gave Shaky a competitive discount for the products they wanted and they agreed to a differentiated solution.*
>
> *On the eve of the closing, Shaky's CEO called a Stable executive not directly involved in the negotiation to ask for a much larger discount and an easing of the differentiation restriction. The Stable executive did not call anyone on the negotiating team before he caved in and gave*

Shaky an additional discount and allowed them to provide the solution to customers without differentiation.

The K&R Deal Forensic

Stable made several mistakes in this situation:

1. Before making a decision, the Stable executive should have called his lead negotiator to get all the facts, not just Shaky's position.
2. Stable's lead negotiator should have done a better job of keeping his executives informed throughout the negotiation process.
3. With either of the first two steps, the Stable executive may have realized that the true problem, the cost of Shaky's infrastructure, would not be solved by his concessions.

Stable's credibility for any future deal was damaged as a result of unprincipled concessions. That is because any value argument the negotiating team now makes will be considered easily overturned by management. When you lose credibility, the next deal with that company will be much more difficult to make. In this case, Shaky will probably continue to negotiate in any future deal until they reach an impasse. Then they will seek out Stable's management again. Wouldn't you? It worked the last time.

Always focus on communicating in both directions: up to management and across to the team.

Lead negotiator can be a job title or can come from any functional area. But we often hear:

- "What if I'm not the lead negotiator?"

- "We are on many deals at the same time, so I rarely lead

negotiations."

- "We don't always lead."

- "I am usually a functional representative, not the lead negotiator."

- "I only have partial responsibility for the deal."

- "I have direct customer responsibility for my products but the lead is usually the customer account manager."

In these situations, it is just as critical to be aware of the negotiator's responsibilities and processes we just described. As a second seat or in a functional role, you want to know what is expected of you and what you need to do to help move the negotiation forward. You may also need to help a lead negotiator, who may not be as well organized as they want to be (or should be) through this process.

Often, a number of different functional roles are involved in a deal, but no lead is assigned. In these situations, you should build consensus at the kickoff meeting regarding who will take the lead for different sessions and subject matter. Sometimes the result is no clear agreement, except to go back to management and get a lead appointed. It is folly to get in front of the other side without having this resolved. (Remember, a divided team is a costly team!)

Being a second seat or functional support in a negotiation is often more difficult than being the lead. That is partly because not speaking "out of turn" is difficult. No lead negotiator will get to the same point the same way as the other team members. As a support person, you may be tempted to jump in to help them. That can interrupt the flow and momentum. It is one more reason why, as a lead or a support person in negotiations, it is critical to prepare with the team.

ADDRESS AND MANAGE INTERNAL CONFLICT

Harvey had the following experience while negotiating a deal in Middle East. See how he resolved the internal conflict:

Our client had put together a strong team to negotiate this deal. This was one of the largest commercial deals in the Middle East for our client; huge commissions were riding on it. The client had a local negotiator who was very talented, but very strong willed. We'll call him "Attila". As a result of Attila's inability to be flexible when necessary, the deal just wasn't getting done. Due to this impasse, I was called in to take over the lead negotiator role.

After arriving, I met with our entire team to assess the situation. As a result, we changed the previous negotiation strategy. It was clearly explained to Attila that I was now in charge and that he would take the important second chair. He agreed. In planning the conduct of our meetings with the customer, I said, "All information would be funneled through me." I reiterated that point a number of times.

On the way to the first meeting, I reminded him that he had an important role—to cover the financials— but that I would take the lead and, again, all communication was to go through me. I repeated this at least four times, and he agreed.

We walked into the meeting room and Attila sat four seats away from me. I motioned him to sit next to me, as is common for the two principals on a large deal team, but he refused. After the introductions, I explained to the other side the roles of the people on my team and my role as lead negotiator.

Within a few minutes after the negotiation began, the client's lead negotiator was speaking. Attila jumped in and interrupted him. I then interrupted Attila after a few words. I politely explained that the client was speaking and not to interrupt him. A few minutes later, he did it again. The third time, I stopped him at the first word. He jumped up and stormed out of the room. I felt compassion for Attila because he did not know my skills well enough, and he had a huge commission riding on the outcome. Nonetheless, I had to manage this internal conflict to keep order and focus. I also had to maintain credibility with the other side and with the rest of my team.

The members of the other side smiled, as we were finally going to move forward.

As Attila saw that negotiations were moving forward, he returned for subsequent sessions and helped close the deal. He got his commission and sends me a local calendar every year.

The K&R Deal Forensic

While this situation was awkward, several good things resulted from the way it was handled to the long-term benefit of all participants.

1. Negotiation rules and expectations were dearly communicated to the entire team prior to the meeting.
2. Those rules were enforced, so credibility with the team and the customer was maintained.
3. When the dissenting team member (Attila) saw that Harvey was negotiating wisely, Attila became his best ally.

Your job as lead negotiator is to set the negotiation rules and reconcile individual and team goals. As the previous story illustrated, you must also negotiate to individual motivations both internally and externally. It is essential that you and your team present a unified voice to the other side. Unless those internal conflicts are resolved, you are not ready to engage the other side in a negotiation. Sometimes, even when you have gotten everyone's agreement, it does not work as planned. You can only try your best and have the courage of your convictions.

Complete the exercise in the workbook to explore this issue:

STOP if you are using the Companion Workbook.

Exercise 12-2: Service with a smile

Scenario 1:

> *You go into a store to buy a refrigerator and several salespeople rush to your side. You feel like a magnet surrounded by metal shavings. What can you conclude about the motivation of the sales force?*

You can conclude that every salesperson wants to sell you a refrigerator because they're probably on commission or get evaluated based on customer service. (Either that, or they are all part owners of the business!) You buy = they get money.

Scenario 2:

> *You go into a store to buy a computer and every salesperson ignores you even though they are not busy. You feel like you have a contagious disease. What can you conclude about the motivation of the sales force?*

You can conclude that the sales force is probably not on commission and doesn't have an ownership interest. Whether you buy or you don't buy makes no difference to them at all.

Even everyday situations like these give you a chance to practice your negotiation skills and examine motivations. Taking the time to deal with each individual motivation and interest makes the deal go faster, and helps you resolve internal and external conflicts.

Explore this issue in greater depth by answering the workbook questions.

STOP if you are using the Companion Workbook.

Exercise 12-3: Negotiation dynamics

Clearly, hard work gains you knowledge, which leads to confidence and earns you credibility, which gives you leverage. Leverage used wisely leads to desired results and successful relationships.

IN THE HOME STRETCH: MOMENTUM TO CLOSURE

Close each negotiation meeting or call this way:

1. Confirm what you have already *agreed* on.

2. Confirm what you *don't agree* on.

3. Confirm on who is going to create a solution to each open issue or address each action item.

4. Agree on when and how the actions and solutions will be delivered.

5. Set the next meeting and its agenda.

Some people don't want to summarize a session because they don't want to bring up the issues that remain unresolved or appear to be closed, but really aren't. They fear having to revisit these issues and spend more time on them at that moment. But it is best to deal with the issues while they are still fresh in everyone's mind. Settle them so you can move on, and if you can't at that moment, at least everyone knows what needs to be addressed. Issues not resolved now will have to be addressed later, when it will be much more difficult and take longer, especially if these issues are not highlighted early. Most important, if you identify the open issues, you can discuss and give appropriate assignments to both sides to move the agenda closer to closing the deal.

Momentum to Closure has an impact on almost all negotiations. It occurs when at least one side in the negotiation perceives that all major issues have been resolved. At this point, the deal takes on momentum that drives people to close even though some of the important issues have not yet been resolved. Instead, these "minor" issues default to their status in the latest draft. Remember this K&R principle: "Terms cost money; someone pays the bill." Once you reach the point of Momentum to Closure, the snowball is rolling down the hill and gathering force as it goes.

Send a letter of thanks to team members at the end of a negotiation.

If team members did an outstanding job, try to get them official company recognition. Not only is that the right thing to do, but they will be available to help when you need them for the next deal—no matter how busy they are.

THE PROOF IS IN THE PUDDING

The pressure is on, and guess who is in the hot seat. Even if it's you and the team is small, a few heads are better than one. It's very difficult for one person to think, interact, observe, listen, and take notes at the same time. We also recognize that many negotiation sessions take place over the phone. Even in these situations, for the more important deals, you can and should enlist others to help you plan and prepare the calls. See Chapter 14 for specific guidelines for dealing with negotiations conducted over the telephone.

Be sure to keep the K&R Negotiation Method™ in mind as you negotiate. As your team debates the issues and works out the best responses, also consider what we've discussed about P&L (patience and listening), credibility, value, and leverage.

As you work out your negotiation strategy, consider these key questions:

- What is at stake?

- What is the value to the other side and how can I quantify that value?

As you learned in previous chapters, there is very rarely only one approach to solving a negotiation problem.

SIGN ON THE DOTTED LINE

IS BIGGER BETTER?

Scenario:

> *You are a line-of-business executive for a mid-sized company (about $500M in annual revenues). Your business needs $2 million this year to accomplish the following:*
>
> - *Hire two staff people at $125,000 each year, fully burdened*
> - *Hire five sales and technical-sales people at $150,000 each a year, fully burdened*
> - *Buy demo equipment and sales material*
> - *Pay for marketing promotions*
> - *Fund the travel budget*
>
> *In order to secure this funding, you must make your case before a committee that includes your sales executive and one member each from the finance, human resources and marketing programs. You believe that this $2 million investment will generate incremental sales of $6 million annually. You are willing to take on that quota. You would like a decision today.*

Not everyone will agree! What persuasive arguments can you make to get the support of each functional representative on the committee? Write your arguments in the workbook.

STOP if you are using the Companion Workbook.

Exercise 12-4: Is bigger better?

Here are some arguments you might have made, with appropriate detail, of course:

1. The initial investment will be realized by year-end.

2. We can hire experienced people or transfer people laterally. No outside hires will be involved, if that is a problem.

3. The lower-skilled positions will be back-filled, which cost less and are easier to fill.

4. The new budgeting and resources for marketing will help increase sales (which is the goal) with an excellent ROI.

5. Overall sales, market share, and profit will be improved within the company's internal measurement objectives.

In all of these, indicate the numbers and projections for a more persuasive approach.

Here is some additional information you could use:

1. Is the company public or private?

2. Are there alternate investment proposals on the table?

3. How large is the company? What are its revenues?

4. Will the investment money have to be borrowed or is cash available?

5. What don't you have that finance is interested in?

6. How real is the $6 million figure? What are the assumptions behind the business case?

7. Is the $6 million a low, medium, or aggressive business case? Can you do better?

8. How much is capital expenditure (capex) versus ongoing operating expense (opex)? [A really sophisticated corollary – will the resulting expense (or SG&A) to revenue (E/R) ratio of the ongoing investment be better than the current company E/R]

9. What is the true ROI from a profit standpoint?

Remember: In most companies, marketing is measured by revenue or market share. In this negotiation, you are trying to personalize each argument to each person. This is how you sell, internally and externally.

THE UNKINDEST CUT OF ALL

Scenario:

> *You are the salesperson responsible for a product we'll call "Dealware." Your company's account executive is responsible for the overall relationship with a customer we'll call "Retail." He is close to closing a $2.6 million multiproduct service solution deal with Retail. Dealware is $288,000 of the total solution after a 40% discount. At $2.6 million, the total deal has an average 35% discount off a list price of $4 million. Your account executive has requested that you discount Dealware an additional 20% off the current offer. This 20% incremental discount would cost you and the account executive $57,600 in revenue and substantially impact profit. It improves the overall discount of the total solution by about 1.5%.*

Dealware provides some unique functionality within the total solution. You can quantify the positive impact of this function on the customer and believe that they are willing to pay for most of it.

In this scenario, you are to develop a rationale that persuades the account executive not to increase the Dealware discount by an additional 20% ($57,600) off the current offer. Ask yourself these questions:

1. Why does the account executive feel the need to discount now?

2. Where does the customer perceive value?

3. What is behind the account executive's request? Why is it in his best interests to avoid eroding the Dealware part of the deal by 20%? The overall deal by 1.5%? Is it really to close faster or is there some other factor at work?

4. Is this request customer-driven? Does more discounting solve the problem?

5. Is the approximate 1.5% additional discount a deal maker or a deal breaker?

Write your rationale in the workbook.

STOP if you are using the Companion Workbook.

Exercise 12-5: The unkindest cut of all

Chapter 12 Negotiator's Responsibilities

	Dealware Discount	Dealware Revenue	Total Solution Revenue (discounted)	Account Exec. Discount Request	Total Solution Discount
Quoted Price	40%	$288K	$2.600M		35%
Revised Price	60%	$230K	$2.542M	60%	36.44%

As mentioned, Dealware provides some unique value to differentiate the solution. So, you have a credible value argument for Dealware. If you now reduce the price of Dealware, what will that say about the value argument for the entire solution? It is in the account executive's interest not to impact the credibility of the value proposition of the entire solution. This is true even if the individual discount for Dealware is invisible to the customer through a bottom-line price. And, unless it is a commodity, is 1.5% going to be the deal breaker? Unless the customer has made a credible value-related argument that would induce the concession on principle, there should be no reason to make the concession. And profit will increase by keeping the price at its current levels. This is a negotiation worth having. The deal is getting eroded, and a smart customer will not stop at 1.5%. Remember, the next deal for Dealware with this customer would start at a discount of 60%. This is an internal sales job.

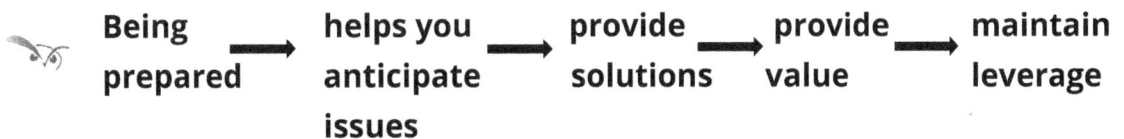

Being prepared → **helps you anticipate issues** → **provide solutions** → **provide value** → **maintain leverage**

K&R NEGOTIATION REMINDERS

Planning, preparation, and teamwork give you and your company the best chance for success. These basics serve to raise skill levels, provide additional knowledge, and help you negotiate wisely. Be sure that the negotiation supports management's goals.

Actions that make your job easier:

- Be patient and listen.

- Understand the motivations of everyone involved.

- Unify your team.

- Quantify your value.

- Restate your value to create leverage.

- Use leverage wisely.

- Maintain credibility.

- Manage the agenda to make progress.

- Break and caucus to change momentum, to gather information, to function as a team.

- Open and close meetings thoughtfully.

- Give the other side responsibilities.

- Watch out for tactics.

- Enjoy the experience!

We will address a few of these actions further in upcoming chapters.

Chapter 12 Negotiator's Responsibilities

WHAT YOU LEARNED IN THIS CHAPTER

- The K&R Negotiation Method™ divides the negotiator's responsibilities this way:

    - Understand and communicate negotiation goals.

    - Establish the negotiation process with the K&R Negotiation Method.

    - Determine the resources (types of skills) that are required.

    - Establish support roles and rules.

    - Earn trust with the team and management.

    - Address and manage internal conflict to bring in the best overall deal.

- Momentum to Closure occurs when at least one side in the negotiation perceives that all major issues have been resolved. At this point the deal takes on momentum that drives people to close even if some issues have not been resolved.

Of course, you also learned the "how" behind those responsibilities.

CHAPTER 13: TACTICS

This amazing story comes from K&R's files:

> Harvey was representing a client in a negotiation. The lead negotiator on the other side had been rude, nasty and downright obnoxious during the process. Things were unpleasant, but Harvey ignored the behavior and refused to let it get a rise out of him. He stayed focused on the merits of the transaction.
>
> Late during one of our meetings, their lead negotiator started to make a point, but then stopped and said, "I'd explain this, but it's so complex that it will just go right over your head, Harvey."
>
> Harvey had several options: He could have replied with equal rudeness. He could have stormed out of the room. He could have done nothing. Instead, he stood up and said, "Feel free to offer your explanation. Now that I'm standing, perhaps it won't go over my head."
>
> Everyone in the room laughed, except the lead negotiator. She got red with embarrassment.

WHAT ARE *TACTICS*?

In the above story, the lead negotiator and the other members of the other side's team believed in negotiating through intimidation and arrogance. They were using **tactics** in an attempt to intimidate Harvey into making a concession. However, they misjudged him: Harvey knew what they were doing and refused to respond to the bait (or in this case, sink to their level). So, what are *tactics*?

In the widest sense, tactics are any action (or non-action). The dictionary defines tactics as:

- an expedient for achieving an end

- the science of using strategy to gain military objective

- skillful maneuvering to achieve a goal

We define **negotiation tactics** as techniques or actions intended to influence a negotiation. For example, great teamwork and coordination to get information useful to your negotiation are types of tactics, as is the use of patience and listening.

We define **gamesmanship** as techniques or actions unrelated to the merits of the transaction that are used to gain an advantage in a negotiation. Thus, gamesmanship is a subset of negotiation tactics. For example, yelling, screaming, or walking out are types of gamesmanship tactics. They are not for everyone, but all negotiators should understand them so they can deal with them.

Gamesmanship as a tactic is usually applied to cause confusion, intimidate, hurry, or improve leverage or momentum. For example:

> *We have a fellow negotiator we'll call "Sami". He liked to employ an especially interesting tactic. On the bigger deals, Sami liked to negotiate into the night—sometimes for 15 hours straight. Around ten or eleven at night, he would call a break and go to the men's room. He would shave and change into a clean, starched white shirt. He would not tell anyone what he had just done. When Sami came back into the room, he looked and felt refreshed. This intimidated the other side. "How come we're so grubby and tired," they wondered, "when Sami looks as fresh as a spring daisy?"*

Sami felt that this tactic gave him a strong psychological advantage in the negotiation. Most people tend to agree. After all, who wants to look and feel shopworn when the person across the table looks wide awake and vibrant? There are many other tactics that negotiators may use. Some are positive and foster relationships, while others are negative and get short-term results, but damage the relationship. We will explore tactics in more depth. But first, complete the activity in the workbook.

STOP if you are using the Companion Workbook.

Exercise 13-1: Sticks and stones may break my bones

In this exercise, you listed the tactics that you believe people use during negotiations to help the negotiators achieve their goals.

LET THE GAMES BEGIN: KINDS OF TACTICS

Look at your worksheet. Do you see any similarities among different entries? As you no doubt noticed, even though the specifics may vary, some of the characteristics of tactics are the same. That's because tactics fall into broad categories. Here are some of the most common categories of tactics (including gamesmanship) used in negotiation:

1. The Stimulus/Response

2. The "Chess Match"

3. The Psychological Stake

4. The Power of "Face"

5. Relieve the Pressure

6. Humor

Let's look at each technique in detail.

Chapter 13: Tactics

STIMULUS/RESPONSE TACTICS

This is a type of tactic geared to getting your way on a specific issue. Here are some examples:

Negative Use of Stimulus/Response Tactic

Mother and Little Goldie are strolling through the mall. Little Goldie is three years old and the apple of her mother's eye. Nonetheless, Mother is determined not to spoil Goldie.

As they walk past a cookie display, Goldie yells out, "I wanna cookie! I wanna coooookie!" Mother replies, "No cookie now, Precious, because we are eating lunch in ten minutes." Born with an instinctive knowledge of the stimulus/response tactic, Little Goldie screams, "COOOOKIEEEE! COOOOKIEEEE NOW!" Naturally, a crowd has gathered.

Can you guess what Mother does? Mother decides that she will teach Little Goldie not to be spoiled—tomorrow. For now, she buys Little Goldie a cookie just to shut her up. Little Goldie's tactic worked brilliantly.

Emotional outbursts, whether screaming, yelling, banging the table or crying, can all be considered stimulus/response-type tactics. They can also be examples of gamesmanship.

Positive Use of Stimulus/Response Tactics

Morris sells printing. Lois buys printing. Morris knows that Lois has a reputation for being a very tough negotiator. "I better stand tough against Lois," he thinks.

Morris decides to take Lois and her negotiating team on a tour of the printing plant. After the tour, Lois says, "My team and I are very impressed with your operation. Your company seems efficient, clean, and well organized. You're a very good businessman." Morris is flattered by the praise. He says to himself, "Gee, Lois is not such a bad person after all."

Chapter 13: Tactics

Morris's resolve has been weakened. Lois' flattery softened him up.

In a way, this stimulus/response tactic is related to giving someone "face". The point is that Lois' tactic worked and seemed to set the stage for an easier negotiation and better relationship.

THE "CHESS MATCH"

This type of tactic sets up the flow of a negotiation in order to gain better results on a series of issues. Here's an example from some coaching we did with a client:

> *Rita is a young executive for Tech Co. She is the lead negotiator on her first major deal with a company that will private label Tech's solution. Her counterpart, Larry, brought in his senior vice president (We'll call him "Ralph") and three other support people for the first set of high-level discussions. After introductions, the following conversation ensued:*
>
> *Ralph: "I want you to guarantee that you will continue to have these products available to us five years from now."*
>
> *Rita: "In this industry, product cycles are 12 to 18 months, so why would you want old product? We will have three generations of new technology by then. This does not seem to make business sense."*
>
> *Ralph: "Maybe I didn't make myself clear. I want you to guarantee that you will continue to have these products available to us five years from now."*
>
> *Rita: "If it makes business sense, I guess we'd have them available. Please tell me again why you would want this guarantee?"*
>
> *Ralph: "If that's how you're going to be, we have nothing left to talk about."*
>
> *With that, Ralph and his team marched out of the room in unison.*

Chapter 13: Tactics

We met with Rita later that day and she was very upset. But we had one advantage—we had no psychological stake in the negotiation. We asked Rita if the deal made business sense for both sides, and she replied, "Yes." We suspected that the walkout was a tactic because it seemed so smoothly executed. We also thought the issue may not be real, but weren't 100% sure. (Was having the same product available five years from now the real problem?)

We suggested to Rita that she wait a day or two to see if the other side cooled down. She should then call the person she had a relationship with—Larry, her counterpart—and try to restart the discussions by exploring why Ralph made the guarantee request. As it turned out, Rita did not have to wait long. The other side's lead negotiator called her the next day and said, "We want to get this deal done, but you saw my S.V.P.—he's wild. You have to work with me and give me the terms and prices we need or Ralph will get involved again." Interestingly, nothing was mentioned about Rita guaranteeing product availability five years from now.

Now we saw the tactic clearly: It was a "good cop/bad cop" routine, with the senior executive being the bad cop no one wanted to deal with. They were trying to set up the flow of negotiations to favor them on subsequent issues. They really didn't care about the five-year product availability guarantee.

Here is another example of gamesmanship under the "chess match" category, from our files:

We were in the position of being the buyers rather than the sellers. They walked into the seller's office for the first time. The sight was amazing: The seller, a 6'5" (196cm) 350-pound man was seated at a desk on a raised platform that looked like a throne. As they walked across the cool tiles, their heels clicking on the surface, I said to my partner, "This is going to be interesting!"

The man was psychologically intimidating because of his size, his posture, and how he had set up the room. He clearly felt he had set up a home-court advantage through a visual display of (negative) leverage.

THE PSYCHOLOGICAL STAKE

This type of tactic exploits a personal, emotional motivation that replaces or impacts logic to get the deal done. The result could be a deal, rather than a *good* deal. A good example is the "Go, Go, CEO" scenario discussed in Chapter 5. Here's another example:

> *Little Billy is failing math. He's not weak in math; he's just lazy. Billy's parents are overwrought and decide to pay him to improve his math grades. "You'll get $50 for an A, $30 for a B, and $20 for a C," they tell him.*

Little Billy will now likely get much better grades in math. But are his parents making a good deal? Although Billy's grade point average will rise, he's now been taught to expect money for good grades. What's to prevent Little Billy from deliberately failing all his courses so he gets paid for *all* his subjects?

Remember that your actions motivate behavior. Is it the right behavior? With a psychological stake, you often don't see clearly enough to figure this out.

We all have psychological stakes in the deals we do, but we must be on guard not to let our emotions interfere with our ability to reason. If that happens, we can be exploited. Here's another example from our files:

> *Our friend Roger called in a panic. Roger, a manager for a K&R client, was in California to sign a deal for sales of $40 million worth of customized systems and services over four years. After six months of negotiations the customer's V.P., Arlene, had called Roger two days earlier and said, "Looks like we're done. The last draft of the contract looks good. Why don't you guys come to California and we'll sign the deal?" Roger had no reason to doubt Arlene.*

Both Roger and his director sent emails to their colleagues and executives announcing that the deal was done. The notes went all the way up the line to the most senior management.

Roger and his director then flew to California and met with Arlene. She said, "I'm really sorry, but my CEO just asked me about our deal and he wanted to see the contract. He focused on the pricing schedule."

The pricing schedule was as follows:

Units	Year 1	Year 2	Years 3	Year 4
1-2,000	$1000	$1000	$1000	$1000
2,000-5000	$975	$975	$975	$975
5,001-10,000	$940	$940	$940	$940
10,001+	$900	$900	$900	$900

Arlene continued: "The CEO claims this is a series of one-year deals; I thought we had a four-year deal. I want to change the schedule so the price would stay at the lowest level for the remainder of the contract after the 10,000-unit level is reached. The impact of this change would be $1.9 million or 4.7%."

Was this a tactic to get a better price or an issue with the agreement structure? How do we find out? What problem are we trying to solve? Roger called us from California and we suggested that they rewrite the grid with the help of a finance person to make it "revenue neutral". So, if the price stayed at the 10,000-unit level, there would be 0% revenue impact overall. That way, we would solve the agreement structure problem at no cost to either side. If the real problem was price, Arlene would have to tell us that the solution didn't work. Unfortunately, in the middle of the phone call with Roger, the director tapped Roger on the shoulder and said, "I signed Arlene's contract. Let's go to lunch."

The K&R Deal Forensic

There are several important teaching points in this scenario.
1. Having spent six months of considerable time and resources on the deal, both sides had a psychological investment to get the deal done.
2. Roger and his director increased their psychological stake by sending internal emails to their associates and executives prior to closing the deal. That made it almost impossible to come back without a deal signed.
3. Psychological and emotional stakes in a negotiation often cloud the rational thought process.

Was it a tactic? We will never know for sure. Psychological stake caused the director to close the deal to his detriment without exploring possibly acceptable alternatives. As is true with any unnerving events in negotiations, if you stay focused on the business merits of the transaction, it shouldn't matter whether it's a tactic or not.

THE POWER OF "FACE"

"Face" is a person's standing in the eyes of others. That means looking good in front of both negotiation teams, peers, management, spouse, family—it is avoiding putting someone in an awkward position that could humiliate or embarrass them. In a positive sense in negotiations, giving someone face makes them feel good and helps form good business relationships. Below is an example:

> *You're in a negotiation with Barbara Brilliant, the new whiz from finance. She passed all four parts of her CPA exam at once, a feat that*

few CPAs accomplish. According to rumor, she was 22 years old at the time. You're thrilled to have her on your team because she is such an impressive number-cruncher, but you feel just a tinge of envy because it all seems to come so easily to her. This is her first big deal with one of the industry leaders. They make an offer and you, as team leader, defer to Barbara to work on the figures. She does it in record time and passes the numbers to you. Only one problem—they're wrong.

You think that you must have made the mistake, so you look at the numbers again. It turns out she's given you bad numbers. Rather than blurting out, "I thought you were ready for the big leagues, but I was wrong," you signal for a break. Then you take Barbara aside and quietly ask her to check the calculations again. This allows her to save face. The rest of the team will never know that she made a mistake. You have earned tremendous trust and credibility with her. When you need her again, she'll be there for you.

Here's another story from our files:

We were representing a buyer of equipment from a Chinese company, OPQ. I was in the second seat, sitting across from OPQ's most senior negotiator, "Yu Jiang". He was serving as a mentor for a much younger team member, Qiang Liu, who was their lead negotiator for this deal. They each had a lot of face riding on this negotiation.

At the end of a long negotiation day, the head of our team said, "Now we will have to address the process for the remainder of our discussions."

Qiang Liu said something important but off the topic. Our lead negotiator jumped on the remark and said tersely, "Let's not talk about that now. We have to address the process." The entire OPQ team was shocked. Yu Jiang was clearly offended. Qiang Liu had lost face not only in front of his team, but also in front of his mentor. This would be serious for anyone, especially to the Chinese, who place a premium on face.

I understood that we would never get anywhere unless we could help the Chinese team save face. So, I turned to Yu Jiang and said, "Qiang Liu makes an excellent point. We will have to address his point before we finish today." This comment pumped air back into their entire team. Yu Jiang supported that comment and praised his lead negotiator for making a good point. Humiliation was avoided, harmony was restored, and the negotiations continued.

Remember: All other things being equal, people do business with people they like—The Power of "Face".

RELIEVE THE PRESSURE

This tactic can be very positive and can be used to let off some steam during tense times. Relieving the pressure comes in two forms: from within the entire negotiation and within a particularly tense meeting. Ways to relieve the pressure include social activities, informal meetings, off-hour activities, and humor. For instance, the teams can have dinner or a drink together, play a round of golf, attend a ball game, or hold a meeting outside on a pleasant day. Changing the environment and getting away from the business conference room can be a very effective way to improve a situation that's gotten testy under the pressure of getting the deal done. Within a meeting, relieving the pressure may involve taking a break or the use of humor.

HUMOR

Always consider your own natural style before you decide how much humor (and what kind) to use. People like Jimmy Fallon, Billy Crystal, and John Oliver are naturally funny and can deliver even the lamest comic lines with a timing that makes them amusing. Other people are not blessed with that gift.

Chapter 13: Tactics

Let's explore the basic rules for using humor to relieve the pressure during a tense negotiation. Keep in mind that some jokes you have heard can be used to fit the situation, while others are impromptu comments that fit the occasion. Both techniques can be effective based on skill, timing, the situation, and your personal style.

Social conventions

To be effective, humor depends on a shared frame of reference. You and your listeners must understand the social conventions underpinning the joke or your efforts will fall flat.

Audience analysis

The humor must fit the audience. Jokes that work during a negotiation with the Midwest Milk Farmers Association likely won't break the tension during a negotiation with a Japanese technology company.

Appropriateness to the occasion

Use humor only when it is appropriate. If you use too much humor during a very stressful meeting when the stakes are extremely high, the humor will likely be viewed as inappropriate.

Personal style

Humor also has to suit your specific style. You may not be comfortable impersonating wild, expressive comedians such as Ricky Gervais or Leslie Jones. Professional negotiation does not involve humor that depends on potentially offensive material, either.

Make a point

When used effectively during a negotiation, humor can do more than just break the tension; it can also make a point. It can support your value statement. The other side might forget the actual joke, but they remember the point it reinforces. Remember the IBID example from

Chapter 13: Tactics

Chapter 2. Here is another example from Mladen's files:

> We were getting close to closing a licensing deal for technology that our client was going to use in an innovative solution for the retail marketplace. The licensor was running out of money at the time and had few alternatives that could get their technology to market in time. They needed this deal badly. Balancing the leverage was the fact that our client firmly believed the licensor's technology was the best alternative and quickest way to get their own solution adopted.
>
> Towards the end of a marathon negotiation session at the licensor's premises, the licensor's CEO, clearly exhausted, started going back to some terms we thought we had agreed to. As we were taking a break, he was about to go to the rest room, and I said, "Where are you going? If you keep going backwards, the building will have run out of toilet paper by the time we are done."
>
> Everyone laughed and we stopped going backwards after we returned from the break.

Ask yourself the following questions before you decide to use humor during a negotiation:

1. Is the joke/pun genuinely funny?

2. Can I comfortably tell this joke?

3. Will the people in the room understand the joke?

4. Will this joke break the tension or otherwise accomplish my purpose?

5. Will the people in the room appreciate and enjoy the joke?

6. Is the joke tasteful?

If you can answer no to any of these questions, play it safe—don't tell it. Follow this iron-clad rule: When in doubt, leave it out.

So, what *can* you say to break the tension, accomplish your purpose, and be funny to most people most of the time? Here are some suggestions to consider.

1. But First, A Little About Me

The safest kinds of humor are jokes that use the safest target—yourself. For example, a child once asked John F. Kennedy how he became a war hero. "It was absolutely involuntary," he answered. "They sank my boat." The linguist S. I. Hayakawa once opened a speech by saying, "I'm going to speak my mind because I have nothing to lose."

The most dangerous humor is the kind that makes fun of other people. Also, it is advisable to avoid these topics:

Religion	Race
Sexuality	Intelligence (unless deprecating your own)
Birthplaces	Disabilities
Ethnicity	Sexual Orientation
Religious leaders	Physical Appearance
Political Leanings	

2. News at Six

Many events that happen to you and to people you know are funnier than anything you could make up or find in a joke book. These events also have the advantage of being fresh, original material, so you won't have to worry that everyone in the room heard the joke from Colbert or Kimmel.

Chapter 13: Tactics

3. Play with Words

A lot of good humor deals with alternate meanings and idiomatic expressions. Baseball legend Yogi Berra can be a great inspiration because his sayings are so silly, yet his jokes work. To break the tension during a tough negotiation, consider these two great "Yogi-isms": "I don't mind being surprised, so long as I know about it beforehand" and "It's *déjà vu* all over again."

However, wordplay does not work well with people for whom English (or the language of the joke) is not the primary language. Be cognizant of your audience.

4. How to Use Humor

If you do decide to use humor, smile and look happy. Your mood will be contagious, making it that much easier to get a laugh and accomplish your purpose.

- As you tell the joke, look at the other side. Pause to focus on individuals you might want to affect. Be sure that a joke won't evoke long-term resentment.

- Keep it short. Use puns. Dragging a joke out can often spoil the humor. You want to relieve the pressure, not bore everyone to death.

- Leave enough time for everyone to enjoy the joke. If you rush the laughter, you undercut the effect you worked so hard to achieve.

- Humor in the negotiation context shouldn't be work. If it doesn't fit or come naturally, don't use it.

- Speak slowly and clearly. Make sure everyone can understand each word of your joke—especially the punch line.

- Sarcasm and irony can backfire, creating sympathy for the

opposition. Think carefully about the outcome before using them.

What should you do if a joke falls flat? Nothing. Never explain a joke. In any group, you're apt to find that some people get the joke and some people don't. Even if the entire group seems baffled, let it go and move on with your point.

IN SUMMARY

So why do people use tactics (or gamesmanship) such as stimulus/response, the "chess match," the power of face, and humor? Because they often work! Tactics can help you—or the other side—gain an advantage in a negotiation. Positive tactics can help cement good relationships that work for both sides. Negative tactics (or gamesmanship) can also help you close a deal, but you should be wary of their effect on a relationship. You need to anticipate the results of these approaches before you use them.

MANAGING TACTICS

Experience shows us that concentrating on the merits of the transaction in a negotiation is the most effective way to close a deal. If the deal offers value, it is good whether or not tactics are used. But it is important to recognize that tactics exist and may be used against you. If a tactic is used well, will you recognize it? Probably not. The best counter to tactics is to stay focused on the merits of the transaction and on the balance of leverage, and you'll do just fine.

You may choose to use tactics yourself. If you do, recognize the inherent risks to your credibility and leverage if the other negotiator realizes that you are using a tactic.

Chapter 13: Tactics

No matter what tactics a negotiator uses, a deal should stand on its merits.

The activity in the workbook involves the use of tactics in a deal. Complete the worksheet now to reinforce what you read.

STOP if you are using the Companion Workbook.

Exercise 13-2: Master of your domain

TACTICS AND GAMESMANSHIP

Many types of tactics can be implemented in a positive or negative manner. The choices are the personal decision of the negotiator. We are not judgmental regarding your decision. Our goal is not to teach you how to use tactics to gain an unfair advantage over the other side. In fact, we rarely initiate negative tactics or engage in gamesmanship. However, we often use tactics in response if the other side has attempted to use them to gain advantages. And even then, our preferred tactic is using humor to defuse the situation and rise above it. We prefer positive tactics that foster relationships rather than gamesmanship. The use of tactics is a personal choice. You decide what works for you.

Here are some additional guidelines for managing tactics:

1. Use the power of your team!

There is no substitute for effective teamwork as a defense to tactics. In fact, teamwork is probably one of the best tactics that everyone can use and respect.

2. Recognition is half the battle.

If you don't identify the tactic, you can't respond to it. Some tactics can be subtle and those using them can be skilled. Also, when you're tired, stressed, or angry, you're more likely to overlook a tactic or have it work effectively against you.

3. Stop and consider.

Once you become aware of tactics, you're apt to see them in every negotiation, even when they're not being used. You may never know if the other side is using tactics—or if they are, whether they intend to. So always ask yourself, "Does it matter?" If not, why bother to respond? In the end, it's the merits of the deal that matter.

4. Use the styles that work best for you.

Remember: "Be yourself. It's who you do best!"

Chapter 13: Tactics

WHAT YOU LEARNED IN THIS CHAPTER

- **Tactics** are techniques or actions intended to influence a negotiation, to gain an advantage.

- Gamesmanship is a technique or action, unrelated to the merits of the transaction, used to gain an advantage in a negotiation. Gamesmanship is a subset of negotiation tactics, usually applied to cause confusion, intimidate, hurry, or decrease the other side's leverage or momentum.

- The most common categories of tactics used in a negotiation include:

  1. The Stimulus/Response

  2. The "Chess Match"

  3. The Psychological Stake 4

  4. The Power of "Face"

  5. Relieve the Pressure

  6. Humor

- No matter what tactics a negotiator uses, a deal should stand on its merits.

- Be yourself. It's who you do best.

CHAPTER 14: INTERNAL AND EXTERNAL NEGOTIATIONS: LOGISTICS AND AGENDAS

Internal negotiations involve cross-functional motivations; external negotiations focus on functional motivations of the other side. They are very similar and key to your persuasion. That's what we will cover in this chapter.

THE INTERNAL NEGOTIATION: CROSS-FUNCTIONAL MOTIVATIONS

What challenges do you think your team will likely encounter on a regular basis?

STOP if you are using the Companion Workbook.

Exercise 14-1: In the hot seat

Here's an interesting negotiation situation:

You are being asked to participate in an ongoing negotiation for your company. Unfortunately, the deal is in trouble before you get involved. You're told: "We need your help! We need you and your team! If we sign this deal, we'll lose millions of dollars a year." You ask, "Why?" They reply, "We need them. We have no leverage. Their prices are too high and they are not backing off!" Half the members of your team say, "We don't want any part of this deal." The rest feel the deal needs to be done. Your instincts tell you that this situation is not as serious as it appears.

In the workbook, list all the challenges you and your team face as you deal with this situation of perceived lack of leverage and a divided team.

We think you'll be dealing with the issues discussed next (arranged in no particular order).

THE TEAMWORK CHALLENGE

Perceived Lack of Leverage

1. One person cannot do it all. Who do we need to help us define our leverage and articulate it?

2. What are examples of leverage that you generally have as a buyer dealing with a supplier? Some are: revenue, volume, market share, profit, reduced costs per unit, credibility in a market, etc. Do any of these apply to this situation?

Divided Team (K&R Principle—A Team Divided Is a Costly Team)

1. Goals are usually driven by measurements. How are the individuals on both sides measured?

2. Goals will vary according to individual roles and responsibilities. What are the goals of our team members? What are their goals?

3. Individual roles and responsibilities cause honest internal conflicts. What are the conflicts?

4. Honest conflict raises important issues. What are the internal issues? What are the external ones?

5. Important issues require resolution. How do we resolve internal issues before we meet with the other side?

6. The issue resolution process can help provide the best solution internally and to the other side. How do we get a unified internal position on the issues?

7. The best internal solutions help the deal work for you. What are they?

8. Does the deal work for *them*? Do we have a business case for them? This bridges us back from the teamwork issue to leverage!

9. Do the proposed solutions support the desired relationship? Does the deal solve the mandatory goals of both sides?

IT'S WHAT'S INSIDE THAT COUNTS

As the following scenario, related to us by a colleague illustrates, sometimes internal negotiations can be more difficult than external ones.

> *It was the beginning of November and we were in negotiations with a company we'll call "Kortell". They had a year-end deadline to close or they would lose their budget for the solution they needed.*
>
> *Our account executive considered that his problem. We said to him, "This is their problem, not ours, so there's no reason to put this problem on our shoulders." The account executive wouldn't listen to our arguments. Instead, he put a lot of pressure on us to make the deal by year-end. He also made us sweeten the deal by offering lower prices as a year-end incentive.*
>
> *The account executive did not need to do any of this because we would have gotten the deal anyway. The value was there for the other side; we had the leverage. Since our account executive didn't understand our leveraged position, he did not have the confidence that this deal would get done by the deadline. As a result, he gave discounts he didn't have to give.*

The K&R Deal Forensic

Several mistakes were made:

1. This team negotiated against themselves and made concessions when none were requested.
2. They threw money at the problem. Money was not the problem.
3. The team should have done a better job explaining the customer's business pressures to the account executive to enable him to recognize that this was truly the customer's problem.
4. The team also needed to effectively communicate to the account executive our value to the customer so he would develop more confidence and patience for our own position.

It is important to plan and prepare thoroughly because goals and roles will vary from deal to deal. Certain rules, however, may not vary. For example, gaining internal agreements before meeting with your external customers is always a good rule to follow. It is essential to obtain management's agreement on how the deal will proceed and what role management will play in the process.

Cross-functional impacts can occur because the deal you are negotiating often affects more than one part of your business. Yet, you don't want a department within your company to raise internal issues late in the process. If this results in changing a position previously agreed to with the other side, it can impact your credibility, erode your Negotiation Capital and cause unnecessary delays. To minimize this possibility, it is important to get everyone who will be impacted involved early.

Understanding these roles and goals for your company will help you and your team to create a negotiation strategy that everyone can agree on. A united internal position offers the best chance to succeed. This is true regardless of the size of your company. The following chart is a general guideline that summarizes the cross-functional motivations behind internal negotiations. Of course, the specifics of your company may not follow this general guideline and should be clearly understood.

Functional Skill	Roles (driven by function)	Goals (driven by measurement)
Lead negotiator	Coordinate team, including Ts & Cs, and prices	Make the best deal for the company
Services	Evaluate task and required resources and alternatives; deliver as promised	Deliverable agreements on profitable engagements
Development	Technical support and feasibility; get technology to market	Cost and time; fund next project
Legal	Review and create Ts and Cs; provide advice on deal	Assess acceptable legal and business risk; facilitate business
Procurement	Assess alternatives	Balance value and quality with reduced cost
Sales/Services Sales	Provide best customer knowledge; know customer politics and strategies; close business deals	Generate (profitable) revenue, customer satisfaction; if service deal closing and bookings (TCV)
Management	Provide support to the lead negotiator as necessary	Close deal with company strategy and metrics
Finance	Provide business modeling; quantify value	Ensure profitability; revenue growth
Marketing	Provide value proposition input (quantify)	Generate volume; support deals

Functional Skill	Roles (driven by function)	Goals (driven by measurement)
Engineering	Evaluate warranty, reliability, parts, quality assurance, credibility	Help provide quality, reliability, good product
Manufacturing	Provide knowledge of product, quality off-the-line, product returns	Help provide low cost, reliable delivery, ordering flexibility
Contracts, Administration and Systems	Provide Ts and Cs that reflect deal implementation of Ts and Cs	Support contract, meet schedules

The lead negotiator may also be a member of one or more of the functional areas. The goals are generally accepted, but may vary from time to time, depending on the specific deal and situation. Individuals reflect their time and place, so it helps to know the specific people involved. That way you can deal with the individual motivations that are driven by experience.

Here's an example from our files of how it should work:

STOP if you are using the Companion Workbook.

Exercise 14-2: Make it work

You are the lead negotiator on a comprehensive-systems deal with a large medical institution, MedCo. Your finance people have told you that the deal needs a minimum 22% EBITDA (earnings before income tax, depreciation and amortization) so your discount maximum is 24%. Your business planning people work with finance to develop a business case for the customer. They pull in the account executive who is most knowledgeable about MedCo and their industry. He argues that this

> *deal will never close below a 30% discount. At 30% the EBITDA on the deal is 16%. Finance is measured on profit. Account executives are measured on 80% revenue and 20% profit.*

What do you do? Write down your actions in your workbook. Then come back and compare your answer with what we did in the text.

This is what we did:

We asked the account executive if he would rather make commissions on a deal at 24% discount or a deal at 30% discount. Of course, the answer was 24%, which would mean more revenue and higher commissions; he was just afraid that, at 24%, the deal would take longer or be lost altogether. But we explained that making a credible offer at 24% would not lose the deal. We always have the option to make another offer to the customer, provided it is presented with good rationale, which, in this case, we have.

Then we challenged the finance person to show us business case data for the customer at a 24% discount. The case was positive, with an acceptable customer ROI for the customer. Some of the assumptions were educated guesses; the customer would have better data in these instances. Then we asked the finance person to make the presentation to the customer together with the account exec. The agreement was that if the customer challenged the assumptions with their own credible data and the result was a deeper discount than 24%, then finance would relent up to a 30% discount. After all, it was not in their interest to lose a profitable deal.

The team made the presentation. After some discussions with the customer regarding the assumptions, we made a principled concession based on an assumption that they corrected, and the deal closed at 26%.

> ### The K&R Deal Forensic
>
> Once internal motivations and goals are understood, logic will usually prevail.
> The fear of losing the deal by the account exec in this case was overcome by the confidence gained through logic and rationale. As a result, when the team is unified and a credible and persuasive business case is discussed with the customer, your chance of a successful outcome increases.

Now let's look at the other side of this coin: the external negotiation.

THE EXTERNAL NEGOTIATION: CUSTOMER FUNCTIONAL MOTIVATIONS

Just as your team members have motivations of their own, so do the members of the customer's team. Start by identifying the key participants and decision makers on the other side. Then think about the company and personal motivations for their actions during a negotiation.

You've explored the challenges *your* team is likely to encounter on a regular basis. What factors will you have to consider as you evaluate the other side's team? Complete the activity in the workbook to explore this issue.

> *A company we'll call "Plasma Plus" makes equipment that analyzes blood and other fluids. They make most of their money on the vials in which the fluids are collected, more so than on the equipment.*
>
> *Here's the risk situation: Our machines were going to be used to process the information and keep the Plasma Plus systems working to analyze the fluids. What happens if our machines cause Plasma Plus's systems to malfunction? It's possible we could be held liable for*

damages caused by incorrect analysis of fluids or by the fluids being spewed around the room due to machine malfunction...

STOP if you are using the Companion Workbook.

Exercise 14-3: The shoe's on the other foot

Below are some questions you are likely to face in this negotiation, as well as in others (not arranged in any specific order):

1. What kinds of risks are possible here?

2. What is the probability of these risks occurring?

3. What liabilities are we expected to bear if these risks occur?

4. Do we insure against these types of risks? If we do, what is the cost?

5. Does the economic return justify these risks and liabilities?

6. What are their company's goals in doing the deal with us?

7. What are the individuals' goals?

8. Will the customer set up conflict between their team and ours as a negotiating strategy to get us to accept more risk?

9. How can I adjust my value arguments to the motivations of the person I am meeting with?

10. Do they have any internal conflicts?

11. Does the deal work for them without us accepting liability for the systems?

12. Do the proposed solutions support the desired relationship?

13. Will the merits of the transaction prevail?

Chapter 14: Internal And External Negotiations: Logistics And Agendas

Typically, you will know less about external motivations than about internal ones, at least at the beginning of a negotiation. Smart negotiators manage information because they realize that the information they provide will be used as leverage against them. The longer the negotiation continues, the more you are likely to learn. But in a shorter negotiation, you may not gather very much information at all. That's why advanced planning is every bit as important externally as internally. Many diverse sources of information can help you determine motivations. Review Chapter 8, "Negotiation Steps: Information Gathering," as you gather the facts.

P&L (patience and listening) are key to determining the set of values you must deliver to get the deal. If some participants on the other side deny or just ignore value arguments, make sure you determine the needs of the ultimate decision maker.

By the way, here's how the situation worked out with Plasma Plus:

> We had an internal meeting with our client's counsel. We believed that any malfunction would likely be caused by "Plasmas Plus" blood analysis software. They were a well-established company with an excellent reputation in the industry. So, we convinced our counsel to take some risk, subject to a reasonable limitation of liability and appropriate indemnification from Plasma Plus. Then we worked as a team to convince Plasma Plus that the value and reliability of our systems justified them giving us the limitation and indemnity protections we sought. We ended up with an $8-10 million/year deal, and it was an excellent deal for them, as well.

The K&R Deal Forensic

This is an example of the internal negotiation that often needs to precede a successful external one. We first had to get our own team to accept some risk, based on the size of the deal and some contract protections. Then we had to make a quantified value argument to convince Plasma Plus to give us the contractual protections we needed. This approach was logical, since we understood both our team's and the customer's functional motivations. Good communication and teamwork by both sides enabled this deal—which lasted for over ten years—to happen.

The following table summarizes most customer's functional motivations related to external negotiations. Understanding these motivations for your specific counterpart will help you and your team create a win-win situation.

Sample Customer Functional Motivations

Their Team	Company Motivations (Based on job role)	Individual Motivations (Based on measurements and career paths)
Executive	Visionary, strategic implementation, competitive advantages, responsible for return to shareholders (Eps)	Desire to keep job, sustain company growth (revenue and or profit), industry KPIs, earn a promotion
Line of Business Advocate for Company	Eliminate the problem, achieve mandatory goals, revenue	Obtain unit results, get recognition, obtain a promotion and pay increase
IT	Provide technical solutions, make these solutions easy to implement, help provide internal customer satisfaction	Maximize technical efficiency, minimize the implementation problems and risks; provide best technology; time and cost

Their Team	Company Motivations (Based on job role)	Individual Motivations (Based on measurements and career paths)
Procurement	Reduce costs by way of low price	Obtain a commission for reducing costs, gain recognition
Finance	ROI, IRR, ROA, profit, cash, profit and loss, make the balance sheet work	Provide good financial performance, become Chief Financial Officer
Legal/Contracts	Provide reasonable protection and minimal risk; provide best Ts and Cs	Implementation (time) of business agreements, Security, Compliance, Career growth
Sales	Grow revenues, increase share, provide customer satisfaction	Found commissions and received promotions
Services	Provide quality support, resource rates discount	Get the job done; generate revenue at low cost.
Engineering	Provide competitive offerings and reliable products	Minimize customer problems and help ensure customer satisfaction at acceptable cost.
Development	Provide product cycle acceleration and cost control	Help adopt new technologies; new and improved products.
Manufacturing	Provide quantity at low unit cost	Limit inventory in order to be a low-cost producer

These are sample motivations representative of many companies we see. Of course, you need to refine this matrix for your specific business partners or customers and their industry/company key performance indicators (KPIs).

A COMPARISON

Let's pull it all together. Internal and external negotiations are similar in many ways. Both can involve and affect cost, profit, product development, resources, time to market, revenue, and other important goals. These are critical areas that help determine your success. So, you should plan and prepare for your internal negotiations in many of the same ways that you prepare for your external negotiations.

While there are many similarities between external and internal aspects of negotiations, there are also differences. As an example, while we advocate open communication internally, communication should be thoughtfully managed externally. For instance, you don't want to give away advantages you've gained through a strong business strategy. You may not want to disclose your product plan too early in the discussion, especially if you don't have the appropriate confidentiality agreement in place. Certain intellectual property, such as your trade secrets, should not be revealed in an external negotiation unless properly protected or licensed. If you do disclose such information, be sure to have a confidentiality or other thoughtful agreement in place.

These comparisons suggest the following similarities between external and internal negotiations:

- Planning and preparation are important.

- A spirit of problem solving applies.

The following chart suggests other points of similarity and difference.

Chapter 14: Internal And External Negotiations: Logistics And Agendas

Internal Negotiations			K&R Concepts	External Negotiations		
Yes	No	Maybe		Yes	No	Maybe
✔			Relationship-oriented	✔		
✔			Motivations	✔		
✔			Objectives	✔		
✔			Requirements	✔		
		✔	Edge (Advantage)	✔		
✔			Affect cost, profit, offerings, and resource	✔		
✔			Credibility applies	✔		
✔			Leverage applies	✔		
✔			Negotiation styles matter	✔		
✔			Persuasive techniques	✔		
		✔	Manage information	✔		
		✔	Protect proprietary information	✔		
	✔		Financial accounting			✔

Internal Negotiations			K&R Concepts	External Negotiations		
Yes	No	Maybe		Yes	No	Maybe
	✔		Financial measurements			✔
✔			Preparation and planning	✔		
✔			Win-win and teamwork	✔		

Conclusion: Process similar; scorecard different. Both affect your bottom and top line results.

YOUR TURN

You've heard a lot about internal and external negotiations in this chapter. Now it's time to apply what you've learned. Complete each activity in the workbook as directed.

STOP if you are using the Companion Workbook.

Exercise 14-4: Roles and goals

Here's how we look at these true-life negotiating scenarios. Compare your responses to ours.

SCENARIO 1: ROLES AND GOALS

You are a sales rep who has done a great sales job. The customer likes your solution because it saves them two headcount per year at

$100,000 each. The solution also increases the productivity of your customer's team by $30,000 in revenue per month. Your customer has evaluated your solution and one competitive solution and confirmed the performance advantage of your approach. Then your customer compared your solution to the way they fulfill this function today. The savings were calculated from the results of analyzing all three alternatives.

1. Placing the order with you

2. Placing the order with your competitor

3. Doing nothing

The price for your solution with a 20% discount is $480,000. Your customer's procurement team, well respected by their boss, declares, "With a 33% discount, I'll issue the P.O. now." The additional 13% discount would reduce the price by about $80,000 to a new price of approximately $400,000. The date is November 30. You would like to close the deal by December 31.

Here are some of the options you might have listed:

1. Take the P.O. (purchase order) now at 33% discount.

2. Make a counter-offer.

3. Ask your customer's procurement team why they are so willing to issue the P.O. now. That way, you can determine if their motivation is something other than cost.

4. Ask your customer's procurement team why they are asking for an additional 13% discount (from 20 to 33%). What is their rationale for the request?

5. Restate the value proposition in terms of delay. The delay costs them $30,000 per month.

6. Grow the deal with services. For example, perhaps you are able to get

their systems up and running faster with some services to help them begin saving money sooner. This may justify the original price and avoid the additional discount.

Which functional skills inside your company can help you?

Finance would be important because this is a financial situation. Your finance representative would play a key role. Your customer account team and their understanding of the customer's industry is also important, as is your technical team, with their understanding of competitive alternatives to your solution.

What's the danger in saying okay to the additional discount? You have sold value all along, and made a credible offer based on that value. If you are willing to drop your price at the slightest pushback from procurement, how credible does the offer you made and the value associated with it look to the customer? Here's the rub: People start doubting their judgment. They also start doubting your value proposition. The customer is putting pressure on you to make an unprincipled concession. Remember: Your goal is to avoid unprincipled concessions and close the deal. If there is a value-related rationale for the discount, such as they will not realize all the value because they can't fully implement the solution, then it may be a principled concession. But you have to find out, and the customer has to "earn" that concession. Adding services to help them implement the software rather than reducing price may be a better way to go because then you are not reducing the perception of value of the solution and getting them used to a lower price.

STOP if you are using the Companion Workbook.

Exercise 14-5: Teamwork and the technical side

Read the following scenario and answer the questions in the workbook.

SCENARIO 2: TEAMWORK AND THE TECHNICAL SIDE

You are having a difficult negotiation with your customer. In order to get close to closing the deal, you believe you must offer some technical services resources for implementation. You hope this offer will allow you to close the deal within thirty days. You make the deal subject to the approval of your technical services team, who aren't present when you make the offer. You did the math and you understand that the offer creates a shortfall of $70,000 to the services team on total service revenues of $280,000. The total revenue for this transaction (including the service revenues of $280,000) is estimated at $1 million.

You are now on your way back to the office to get agreement with your services team on the offer you made regarding their resources. Time is of the essence—you would like to get the technical team's approval, go back to the customer and get the deal signed right away.

Your alternatives include:

1. Make service revenues whole by discounting the non-service components of the deal.

2. Approach the technical team to reduce their price on services by some or all of the $70,000.

3. Ask the customer for some "in kind" commitment, such as being a test site for your next product version or being a reference.

4. Offer any combination of the above.

5. Speak with the technical services team. Then go back to the customer and say, "I spoke with the services team. They would like to meet with you to discuss what it is you are trying to accomplish and how their services will help you accomplish it." Then that discussion will point out the additional value technical services provides to the customer. It may enable you to get back some or all of the $70,000.

Which alternative has the best chance of success?

Option 5 is probably your best bet. No matter how you dance around the issue, you are usually better off articulating value to the customer. If you go around making commitments for the technical services people (or anyone else, for that matter) without their prior approval, they will usually get upset and may not support you when you need them. The best way to solve issues like this one is with teamwork and communication before you meet with the customer. However, based on what you've already told the customer, there is a good chance you will wind up with Option 4.

If you could do the negotiation over, what would you do differently?

You would probably obtain pre-authorization from the technical services organization to use some of their resources to close the deal. Or you might set up the customer meeting with your services team present to articulate the value of their offering. Then a principled concession of adding technical service resources to help the customers get the value for their purchase could be used to seal the deal quickly. In either case, before you make such a concession, you want to make sure you and your customer have the right approvals in order. You want this concession to close the deal and not give the customer the opportunity to ask for additional concessions. Ask for deal signatures immediately.

LOGISTICS

In today's fast-paced business world, many deals are done over the telephone or Internet. When should you use these means of communicating and when do you hold face-to-face meetings? When you do meet, where do you meet? Such logistical questions should be consciously addressed or they can have negative consequences on the negotiation.

LOCATION

Some people prefer to negotiate in a neutral location. This offers the advantages of minimizing distractions and causing both parties to invest in the process. However, using only neutral locations loses the benefits that come from being at either party's location.

What are some benefits of having the other side visit you? First, you can show off your organization, maybe give the other team a tour of the facility or have them meet key personnel in your company. You are also on your "home turf", so you don't need to deal with time changes or overnight accommodations. Further, you have full access to the support of your staff, office files, and any resources necessary to discuss the items on the agenda. Negotiating in your own location also forces the other side to spend their money, time, and resources, which increases their psychological stake in getting the deal done with you.

On the other hand, at least once during a negotiation (especially one involving use of resources and facilities and considerable investment), it's a good idea to visit the other side's facilities and get a tour. As discussed earlier, a personal visit is often worth a thousand words. Visiting the other side can also be less distracting for you. You can concentrate on the deal, and on gathering information that you can only discover when you visit them.

TELEPHONE NEGOTIATIONS

In some ways, negotiations by telephone are more efficient, less costly, and less time consuming than face-to-face negotiations. You can negotiate on the telephone from your office without travel and with full access to your files, the staff, and any expertise you may require to address the items on the agenda. Selecting a location for the negotiation is not an issue.

In your office, you may also have access to Internet resources that can provide you with instant information on subjects that come up. (Of course, today that may be true anywhere). For example, if a customer mentions a competitor, you can access information about that competitor while you are doing the negotiation. Unfortunately, this multitasking can be very distracting and cause productivity or the agenda to slip.

Telephone or conference negotiations have additional drawbacks and risks. For one, if you are negotiating with people you have never met, it's harder to interpret anything beyond the literal meaning of what they say. While voice inflection comes through over the phone, without body language it is difficult to grasp the full meaning of a person's conversation. Without meeting the person face to face, it is difficult to understand the subtext of their speech, though video conferencing can help in this regard.

Once we have met someone in person, we can usually envision their body language from their tone of voice, provided that we both speak the same primary (main) language. But even then, you still can't see their reactions to what you are saying over the phone.

It's easier to get distracted when on the phone. Maybe someone walks in and the other side puts the phone on mute to talk with them, or you get an email you feel compelled to answer.

One of the biggest problems is when your team is not all together during a conference call: Either everyone is on their own phone or some people are in the room with the customer and some are not. It is difficult to manage a team this way, and almost impossible to stop someone once they start misspeaking.

Note: That's a good general reminder with respect to all speech. If you have any doubt about what you are going to say, don't say it. You can always say it later, but if you have already said it, it's too late to take it back.

Also, don't think you can control speech with instant messaging (IM). The conversation often moves too fast to be controlled over IM.

Here are a few pointers for your telephone negotiations:

1. First, as with all negotiations, listen very carefully. That means don't respond to unrelated email, cell phone messages, and any other electronic communications while you are in the conversation. Don't multitask unless it's directly relevant to the negotiation. Also, ask not to be interrupted by people outside the call.

2. Silence is key, so get the other side to do most of the talking while you ask questions.

3. Try to have your team all together. If you are at all close by, a little traveling inconvenience is worth the effort to have everyone on your team in the same room.

4. Whether in the same room or not, get agreement with your team on the "rules of engagement".

5. As always, confirm in writing what has been covered during the conversation to avoid misunderstandings later.

Here's a remarkable story of a phone negotiation from Mladen's files. He was working with a software client we'll call "Honorable", negotiating a licensing and development deal with a major industry player we'll call "Disjointed". The four representatives from Disjointed were in three separate offices in the same complex. The Honorable team all sat in one conference room. As Mladen tells it:

> We needed to discuss some things internally, so we said, "We are going to put you on mute." They said "OK," but as we put them on mute, they continued talking, discussing what they might be willing to pay for the development effort. Within a few seconds we realized that they might not be aware we could hear them. Being Honorable and not wanting to eavesdrop, we turned off mute and said, "You understand that we are on mute, not on hold, so we can hear you." Again, they said, "OK." We put them back on mute and this time we heard one of their team members say, "This development is critical, so I am willing to go up to $1 million." By the time we were able to tell them we would hang up and call them back later, they had spilled the beans on the critical point that made the negotiation a lot easier for us.

INTERNET/EMAIL NEGOTIATIONS

As with telephone negotiations or face-to-face meetings at your location, negotiations by email save money on travel and reduce worries about location and office logistics while providing you with the advantages of access to your office files and resources. Although some people would argue that email also saves time, often negotiating by email can take longer than telephone and in-person negotiations. It's been our experience that taking even a single moderately complex issue to conclusion by email takes several days of exchanges. Conducting the same negotiation by phone or in person can take a single conversation.

There are a lot of things to be cautious about in email exchanges.

Email is a flat medium. There is no room to shape the interpretation of your message by someone else. If you don't write clearly, even the best intended messages can be misinterpreted. Simple, seemingly clear responses are subject to interpretation.

Here's an example:

> *While traveling overseas, Mladen responded to a series of well-thought-out suggestions on a deal from one of K&R's associates by sending an email that simply said, "Good thoughts." Given such a simple, terse response, the K&R associate thought she had wasted her time. A day later, Mladen replied to the customer with a much longer and more carefully-worded email with all of the associate's suggestions incorporated.*

Written statements tend to be taken more seriously than oral conversation. As a result, some messages not intended as commitments can be taken as binding just because they are in writing. They may come back to haunt you in the future, especially if they are interpreted as negative.

Even more so than with telephone conversations, with email, you lose the benefit of reading the other side's body language. That's one of the reasons why it's difficult to build trust and confidence and to establish a relationship by email alone.

Writing email requires even more attention than verbal communication, which can be corrected on the spot. You can never take speech back, but you do have the chance to correct any misunderstandings. Yet, when we reply to email, we value speed over clarity, so we often fail to scrutinize what we write.

Keep in mind that when answering email, you are facing a screen, but you are corresponding with people. Acting quickly is easier; yet. the require-ments of negotiating by email are no different from those of face-to-face or telephone negotiations. All of these negotiations require preparation, planning, patience, understanding people, knowing the needs of the other party, persuasive skills, and problem-solving capabilities. As with responding to letters, email requires you to give each message full consideration, including assessing how your message can be interpreted or misinterpreted. Some ecommerce tutorials recommend waiting up to 48 hours before replying to an important email message. While we can't always wait that long, it may help to print a draft of your message and let someone else read it (teamwork) to make sure you are on firm ground. You may be able to buy time by sending an interim message. That enables you to maintain communication while taking time to prepare for negotiation.

Overall, negotiating by email is difficult and should be limited to document exchanges, information exchanges, clarifying issues otherwise discussed, or finalizing specific clauses in agreements. And be careful to make sure your documents are not altered by accident or intentionally.

Of course, email can be a good medium for arranging meetings, exchanging agendas, introducing participants, or confirming what has been discussed. But even for small companies, despite the benefits of negotiating from your office by phone or email, it's often easiest to cover the largest number of important points in a negotiation in person (or by phone). That's because you have the attention and commitment of both sides when they take the time to be there.

Visiting or calling your customers or business partners after the negotiation has ended is important for maintaining personal relationships and for keeping a pulse on their business environment. That goes for small companies and large companies alike!

MANAGING THE AGENDA

In our view, too much is made of "agenda control", It's nice to control the agenda, but both sides in a negotiation have issues that need to be addressed, so control is not as important as _managing_ the agenda.

You'll manage the agenda on two levels: *macro* and *micro.*

- Macro = the big picture

- Micro = the details and execution

ON THE MACRO LEVEL...

The agenda on the macro level refers to the entire negotiating process, including both internal and external discussions, the timing of document creation and exchange, and meetings. Remember that participants' motivations are changeable, not static. So, you must revisit your position as the process moves forward, as well as when team members learn more and reassess the situation.

If you have excellent value creation and qualifications that solve the other side's problems as they emerge or vice versa, you may be able to accelerate the process. That's why getting and confirming agreement on timing is important. If you believe you can accelerate the process to closing, you must engage the right resources on both sides to make decisions.

Be sure you have knowledgeable resources from your side available at the right times in the process. That way, decisions can be made and the agenda will move forward to the closing date.

The following checklist illustrates the macro agenda internally and externally.

Develop negotiation strategy <u>internally</u>

- ✓ Develop a "Leverage Profile" for the deal

- ✓ Identify your company's resources and motivations

- ✓ Identify customer needs and weaknesses

✓ Identify customer strengths or their other leverage points

✓ Identify ways to address the customer's leverage points

✓ Identify your unique values that give you leverage

✓ Quantify value in the context of customer weaknesses

✓ Develop target timeline, the start and end dates

✓ Establish your company's unique criteria

✓ Establish the NSR strategy

✓ Determine the role(s) of participants and what role management will have in the process

✓ Confirm your approval process

Confirm negotiation strategy externally

✓ Establish start and end dates

✓ Provide a compelling reason to make decisions

✓ Understand customer alternatives

✓ See if there is any advantage to delaying (for the other side)

✓ Understand customer decision-making process

✓ Confirm that this deal is worthwhile for the customer and for you

✓ Understand their approval process

We like to manage the macro agenda by working backward. We start with the date on which the customer wants to close and work back through the steps that need to occur in order to make that date. For example, if we want to close by December 31, we put in a target date of November 15 to allow contingency time for unpredictable events. Then we work back from November 15. Usually, the last set of approvals from the other side is from finance and legal counsel. That means that around October 15, we will need a close-to-final business case from our finance—and sometimes business development—people. Make sure you obtain all the necessary internal sign-offs.

Another way to help you manage the agenda is to think about habits developed over time. For example, if you are in a quarter-to-quarter business, particularly sales, you likely feel very pressured at the end of every quarter. The pace of your activity increases exponentially. After the quarter is over, you relax and slow down. Do your counterparts from the customer side feel the same pressure at the end of a quarter or year? "Well, they are not on quota", you might say. But think about this: If you are in procurement, finance, IT management, or even legal, wouldn't you feel pressure and overload if you knew that at the end of every quarter or year, 100 different vendors would descend upon you, trying to get your time to strike a deal? Turn this situation around. Instead of relaxing when the quarter is over, schedule preliminary meetings with your counterparts. That way, they give you their full attention when they are not overwhelmed. They can listen better to your value proposition. It doesn't mean that you will make the deal earlier—but you may. And when it comes to the end of the quarter, who are they most likely to give their time to? You—who they have met before and feel comfortable with—or some other vendor they have never met?

ON THE MICRO LEVEL...

The agenda on the micro level includes the activities surrounding a single negotiation meeting. The agenda starts before the actual meeting and ends after it. Preparation and follow-up are critical to getting closer to closing. Fail to prepare, fail to confirm at start and end, or fail to follow up -- and you lose your momentum. These kinds of "expectation" failures create the opportunity for misunderstanding. They also waste time and resources, both internally and externally. And they delay closing.

The following graphic illustrates a K&R micro agenda management process. The goal of the micro agenda is for every interaction to help the macro agenda move closer to closing.

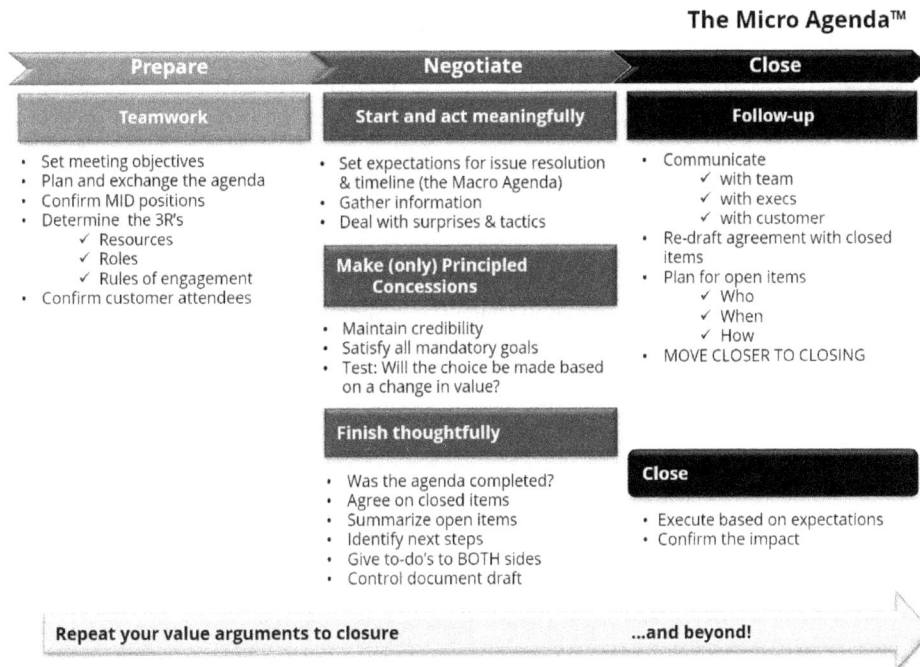

The Micro Agenda™

Prepare	Negotiate	Close
Teamwork	**Start and act meaningfully**	**Follow-up**
• Set meeting objectives • Plan and exchange the agenda • Confirm MID positions • Determine the 3R's ✓ Resources ✓ Roles ✓ Rules of engagement • Confirm customer attendees	• Set expectations for issue resolution & timeline (the Macro Agenda) • Gather information • Deal with surprises & tactics **Make (only) Principled Concessions** • Maintain credibility • Satisfy all mandatory goals • Test: Will the choice be made based on a change in value? **Finish thoughtfully** • Was the agenda completed? • Agree on closed items • Summarize open items • Identify next steps • Give to-do's to BOTH sides • Control document draft	• Communicate ✓ with team ✓ with execs ✓ with customer • Re-draft agreement with closed items • Plan for open items ✓ Who ✓ When ✓ How • MOVE CLOSER TO CLOSING **Close** • Execute based on expectations • Confirm the impact

Repeat your value arguments to closure   **...and beyond!**

Figure 1: The Micro Agenda™

We have discussed many of these subjects in previous chapters. Two of the most important aspects of micro agenda management are the opening and closing of meetings or interactions (if by phone).

Opening begins before the meeting starts. It includes setting the expected subject matter to be discussed and the appropriate attendees. This is when we set expectations for both sides. We ought to have the right resources available to fulfill those expectations. To facilitate this, sending a simple agenda before the meeting is appropriate.

In closing meetings, there are two tendencies: One is that we do not want to summarize open items, particularly if the meeting was a good one. We tend not to want to raise issues on which we disagree. However, those issues do not go away on their own. Raising them sooner is critical to defining a path and timeline for resolution. The second tendency is apparent with salespeople or services professionals, who want to please the customer and solve problems. When open items are raised, they often take on the responsibility to solve them all. Don't do this! Be selective, but give your customer some tasks to complete. Let them be responsible for doing some of the homework. Both parties should be spending time, money, and resources on the deal. Otherwise, it is only you who has the increasing psychological stake. Let them feel ownership for getting this deal done with you.

WHAT YOU LEARNED IN THIS CHAPTER

- Internal negotiations involve cross-functional motivations.

- External negotiations focus on customer functional motivations.

- Internal negotiations can be more difficult than external deals.

- Plan and prepare thoroughly because goals and roles will vary by deal.

- Identify key participants and decision makers on the other side; think about their company and personal motivations for their actions during a negotiation.

- P&L (patience and listening) are the keys to determining the set of values you must deliver to get the deal.

- Good teamwork is a strong asset in any negotiation.

- Set up your macro agenda early and work the timetable backward.

- Ensure that each interaction moves the process closer to closing.

- For each meeting to be productive, set expectations early and have both sides' resources available.

- Close each interaction by summarizing open items and identifying a path to resolution.

- Do not take on all the responsibilities for getting the deal done; give the other side tasks to perform.

- Understand and deal with the risks of telephone negotiations.

- Be careful when communicating by email in negotiations.

CHAPTER 15: THE IMPORTANCE OF TEAMWORK

Consider this story from our files. It took place while we were negotiating a major deal with a group of highly-trained technical people.

SIX PEOPLE, ONE BRAIN

Our client, "Tech" had a negotiation team composed entirely of very experienced techies. This team of smart senior developers had no one on their team from sales, finance, procurement, business development, or any other functional organization.

Before we came on the scene, they were involved in a negotiation to create some industry standards with another industry leader we'll call "Balanced". They had a balanced team that included some techies as well as some nontechnical sales, legal, and business development staff.

What was the key difference in the makeup of the two teams?

Balanced had a cross-functional team comprised of people from different business disciplines. Before the detail work began, Balanced and Tech had already agreed on a 70/30 split of the funding (paid for by a third-party consortium), with 70% to be paid to Balanced. At 30%, Tech broke even. The team did not see this as a problem. Keep this split in mind as you read on.

The negotiations were going well when they suddenly hit a wall. The problem arose over the issue of ownership of the finished product.

Balanced wanted sole ownership. Tech wasn't sure why but felt that relinquishing ownership would be a problem for them down the road. They called us to see if we could get the talks back on track. What factor do you think caused them to have problems negotiating this deal? The problem was with the structure of Tech's team. The team had great technical experience but lacked experience in key business areas. People tend to gravitate toward those they know, like, and respect. The techies knew a lot of smart, talented people just like them...yes, you guessed it: other techies. Tech felt good about their team because they were all comfortable together and, they all spoke the same language: "Tech Speak"!

So, the techies didn't feel any need to create a cross-functional team. They were singularly focused on establishing an industry tools standard. As far as they were concerned, ownership, profit, distribution, and future rights were mysterious. They had not created a statement of work; they had not done any preparation on the financial work. Future rights to the product hadn't occurred to them. Because the team wasn't balanced, key issues had been ignored, misunderstood, and neglected.

BALANCE TEAMS

The moral of this story? Effective teams must be balanced, even if only a few people are involved. You want people who think differently from you— to challenge you and bring a different perspective. Having cross-functional representation is an asset.

Here's a list of some key issues that Tech had difficulty grasping prior to working with their business team:

1. Tech believed the sole problem was ownership; Balanced wanted sole ownership of a product that was being jointly developed.

2. Tech maintained that they had mixed feelings about making money. Their sole goal was to enhance developer productivity and create a tools standard. They firmly believed that making money and improving developer productivity were mutually exclusive. We pointed out that these goals could (and often do) exist in the same

universe.

3. Tech had agreed to a financial split before they did the detail work. Here's what the detail work looked like:

Item	Tech	Balanced
Code contribution	90%	10%
Resource contribution	80%	20%
Financial Split as proposed by Balanced	30%	70%
Equitable Financial Split based soley on code and resource contributor	85%	15%

Here's how the deal worked out: Tech was on its way to making a lousy deal. We worked out a much better one. Creating a cross-functional team and gathering information worked in our favor. This knowledge gave us the ability to make rational, acceptable arguments to the other side. The deal was restructured to allow Tech to get equal ownership rights and a reasonable share of revenue and profit based on their investment in code and resource.

IS A BIGGER TEAM A BETTER TEAM?

Let's consider the size of the team. Do you want to assemble a big team or a small team? Do good things come in small packages? Is bigger better? Using the workbook, fill the chart with advantages and disadvantages of big and small negotiation teams.

Chapter 15: The Importance of Teamwork

STOP if you are using the Companion Workbook.

Exercise 15-1: Big negotiation team/small negotiation team

As is often the case, the answer to this is: It depends. The factors to consider are the importance of the deal, the size of the company involved, the resources available, the subject matter to be covered, the complexity of the issues, and much more.

On one hand, a large negotiation team can create an impression of size and importance. On the other hand, it can also create the impression of inefficiency, conflict, and confusion. Plus, the more people you have actually meeting with the other side, the greater the risk that someone on your team will misspeak. Our advice? Create a balanced team with the resources that can best deal with the issues on the table. The extended team should have all the resources represented to address the macro agenda. That doesn't mean that all these people are actually meeting with the other side. In fact, as we have seen, many valuable team members never negotiate directly with the other side.

Initial deals can turn into business relationships that last for many years. Requirements will change; new products and services will be announced. You want the relationship to survive, so be sure to put people in place that have the capability to stay involved with the other side after the deal is signed.

The best minds for the deal will help ensure that the right deal gets done for both sides—and especially for your side.

In today's complex business environment, a single person usually cannot handle all aspects of the negotiation process. How likely is it that one person has all the technical, sales, financial, managerial, legal, services, and procurement skills that might be required to close even a relatively simple deal? And how could they also have the negotiation skills, contract writing experience, and terms and conditions knowledge required? And then they would need to be able to write, negotiate, listen, think, and observe everything that is taking place during the negotiation itself. We wouldn't want to leave an important deal in the hands of just one person. And we wouldn't want to be that person with all the pressure on our shoulders! That does not mean there is no person in charge—that responsibility must land somewhere, or else confusion will reign.

But what if you are a small company or business unit with limited resources? If the deal is important enough, get external resources. We do this all the time. We have clients that are small companies, but on the more important deals, they have every function involved and use external resources such as K&R or other experts to discuss the issues and help negotiate.

Regardless of size, your negotiation team must stay united. The other side will often try to divide you, so take measures beforehand to help everyone work together. We suggest you set up rules for team communication, the "rules of engagement", in advance of any negotiation.

UNITED WE STAND, DIVIDED WE FALL

Any negotiation can be a lengthy process. The lead negotiator may not be there for every phone call, on every subject, every time. As the lead negotiator, you will often need to delegate to responsible team members. But remember these guidelines:

- Don't delegate without inspection and follow-up.

- Don't delegate without prior coaching.

- Set negotiation parameters for those to whom you delegate.

The role of a support person is one of the most important, yet difficult, roles in any negotiation. As a support person, it is difficult not to break in when the lead negotiator is not making a point the same way that you would. However, no two people get to the same point the same way— even if they trained together! That's why team preparation in advance of each negotiating session is crucial. It builds confidence and trust.

A lead negotiator who has earned the confidence of team members has a much better chance of minimizing interruptions. That's because team members will trust that the lead negotiator will get to the point even though the path may be different from what they might have taken.

Decide how you would handle the following negotiation. As you think about the issues (especially leverage and the Negotiation Success Range), consider how you can work as a team to make the deal a success.

> *We have two companies, Kumquat and Incline. Kumquat is a well-established international company, while Incline is small, with only a North American presence. With Incline's help, the developers at Kumquat created a specialized computer board that worked only with the routers that Incline built. Unfortunately, the Kumquat engineers developed the computer board before they negotiated the right to buy Incline's product at a discount. And they needed the product to provide their customer solutions. Kumquat's people had spent a great deal of money and time developing this board. The Kumquat engineers realized: "We did this backward."*
>
> *Kumquat had given Incline a large amount of leverage. Incline used their leverage to negotiate a deal to provide Incline's product (the routers) to Kumquat. As the negotiation stood, Kumquat would lose money on every Incline router they sold. Incline clearly was not using leverage wisely, but held the belief that Kumquat would still do the deal*

because they needed the routers for their solutions which could "hide" the loss.

You negotiate for Kumquat. You are called in to get a better deal for them. How will you and your team do it? Write your strategy and solution in the workbook.

STOP if you are using the Companion Workbook.

Exercise 15-2: If you build it, they will come

Here's one possible response: Work with your team to establish where you may still have leverage, then take the time to use that leverage wisely. So, let's examine your leverage. As Kumquat, you:

1. Deal on an international scope, which can provide Incline with added market opportunity

2. Provide huge potential revenue and volumes for Incline, provided you are motivated to sell

3. Can give a small company instant credibility

4. Can issue a joint press release to further Incline's credibility

5. Pay your bills on time (and the checks always clear!)

6. May be forced to walk away; the deal can be a win-win, a win-lose or lose-lose

At this late stage, the only way you can change the status of the deal is with persistence and patience. That means you need to make a repetitive value argument focused on your leverage points and have the patience to obtain movement from the other side. Ultimately, Incline should do the right thing for both parties because it is in Incline's best interests. It's up to you to point that out and motivate the right behavior. **Articulate**

Value! (In real life, they did and the deal got done as a win-win, but Incline still got a better deal because of the early mistake by Kumquat!).

The K&R Deal Forensic

The Kumquat team made several mistakes:

- The first was in not understanding how leverage would shift to Incline as Kumquat continued to invest money, resources, and time in the development of the computer board, a solution that was dependent on Incline.
- The second was in not recognizing their own leverage.

In real life, K&R was able to help Kumquat in several ways:

1. We had no psychological stake and could analyze the circumstances more clearly.
2. Our analysis focused on the leverage points, including the worldwide volumes and revenues Kumquat would provide to Incline.
3. Kumquat had a large psychological stake in the deal. By getting the Kumquat team to recognize their own leverage, and through persistence and patience in articulating that leverage, we were able to get a satisfactory deal for both sides.

Close faster! Do teamwork early.

TEAMWORK: THE BEST TACTIC

Ever hear this joke?

> *Two kids were trying to figure out what game to play. One said, "Let's play doctor."*
>
> *"Good idea," said the other. "You operate, and I'll sue."*

It's a silly joke, but it does illustrate one way that teamwork pays off. Here are our five guidelines for effective teamwork:

1. Recognize varied motivations, both internal and external.

2. Debate internally; unify externally.

3. Build two-way trust and responsibility with management.

4. Create a true team in which everyone contributes expertise.

5. Work as a team; communicate openly within the team.

Let's look at each guideline in more detail.

1. RECOGNIZE VARIED MOTIVATIONS, BOTH INTERNAL AND EXTERNAL

Everyone has specific reasons for acting as they do. For example, the other side may want to stall negotiations if they think you need to book revenue now. Their goal is to force you into concessions due to time pressure. To deal with this particular motivation, the time allotted for a negotiation should not be open-ended. Set up a timeline for the negotiation process and work backward (remember macro-agenda management). Review it with your team members, including management, to get their concurrence and their availability. Then get the approval of the other side to the timeline.

The best motivation is self-interest. If you can quantify value so the other side loses more by stalling than by deciding, or gains more by acting, they will decide to go ahead with the transaction.

The first step in dealing with motivations is recognizing them.

2. DEBATE INTERNALLY; UNIFY EXTERNALLY.

Your team should understand the Six Principles and the concepts behind them. Here's a quick refresher from Chapter 6.

1. Get M.O.R.E.—Preparation is key to a winning negotiation.

2. Protect your weaknesses; utilize theirs.

3. A team divided is a costly team.

4. Concessions easily given appear of little value.

5. Negotiation is a continuous process.

6. Terms cost money; someone pays the bill.

Your team must understand the goals of your negotiation. And even though it can be hard to accept, everyone must be aware that winning is uneven. However, while winning is uneven, each side has to win enough to go forward with the deal. To maintain a relationship, each side needs to feel good about the deal. That's why it is important to have an internal discussion about *how* you win. Debating internally is important because it gets the best ideas to the table from a variety of people and interests. The last thing you would want is six people with one brain like our Tech friends! But these debates should be internal only. When you interact with the other side, you must present a unified front to them. Keep the squabbles and differences of opinion behind closed doors.

3. BUILD TWO-WAY TRUST AND RESPONSIBILITY WITH MANAGEMENT.

Even though there is often an "us versus them" relationship with management, management is an asset to the team. This relationship has to be handled properly. Start by realizing that you are all on the same side and want the same thing—to get the best deal for the company. Complete the activity in the workbook to brainstorm some ways that you as lead negotiator can build trust for your team with management.

STOP if you are using the Companion Workbook.

Exercise 15-3: Establish the ground rules

4. CREATE A TRUE TEAM IN WHICH EVERYONE CONTRIBUTES EXPERTISE.

Most deals in the technology industry are too complex to be done by just one person. Too much happens at the same time. As a result,

- Multiple skills are required.
- The art and science of negotiation is required.

An individual cannot be an expert in all subjects, but internal skills are available to make the team collectively "an expert". Use this combined expertise wisely.

When your team meets to prepare for a negotiation, you may wish to assign the following roles to your team members. This can help ensure that everyone contributes their skills as required by the subject matter.

As you study the chart that follows, recognize that roles can change depending on the issues being discussed. *All* roles can change, even that of the lead negotiator. In smaller organizations, a few of the roles are often filled by a single person.

Roles:	Goals/Measures:
Lead negotiator	Bring in the best overall deal
Finance	Profit
Marketing/Sales	Revenue, market share, growth
Development	Cost and asset control; funding for new products
Services	Bookings, pre-tax income, revenue
Other Roles Based on Need:	
Manufacturing	Quality, cost effectiveness
Contracts	Competitive terms and conclusions
Engineering	Specifications, reliability
Procurement	Balance cost and quality
Legal	Risk assessment, reasonableness, probabilities
Management	Parameters for maximum revenue and profit
Administration/Systems	Automation cost and efficiency

5. WORK AS A TEAM: COMMUNICATE OPENLY

As a general principle, the lead negotiator should share all information with the team. This approach will help team members provide more valuable input into the deal. When K&R leaders negotiate, everything they know is communicated to their team openly and honestly. They build trust within the team and, as a result, rarely get undermined by team members or management.

Communication with team members should be:

1. Open

2. Timely

3. Vertical

4. Horizontal

Remember what you learned about communication in Chapter 3: It's a two- way process. If you are in a support position and the lead negotiator doesn't fill you in on everything you need to know to do your best, don't be shy about asking for the information you are missing. If you don't ask, you don't get.

PULL IT TOGETHER

Let's review the essence of good teamwork:

1. Recognize varied motivations, both internal and external.

2. Debate internally; unify externally.

3. Build two-way trust and responsibility with management.

4. Create a true team in which everyone contributes expertise.

5. Work as a team: Communicate openly.

Use the following to help integrate successful teamwork with planning, preparation, and the Six Principles™:

Chapter 15: The Importance of Teamwork

	YES	No
Is there executive and team agreement on the goals of this transaction?	☐	☐
Is your key executive in agreement with your negotiation strategy?	☐	☐
Are cross-functional executives in agreement with your negotiation strategy?	☐	☐
Do you need an internal escalation process to reolve potential conflicting issues?	☐	☐
Are your team members in agreement with your negotiating process?	☐	☐
Management	☐	☐
Product support	☐	☐
Development	☐	☐
Engineering	☐	☐
Manufacturing	☐	☐
Marketing/Sales	☐	☐
Services	☐	☐
Project Management	☐	☐
Service Delivery	☐	☐
Finance	☐	☐

	YES	NO
Legal	☐	☐
Administrative	☐	☐
Procurement	☐	☐
Other	☐	☐
Are you using a draft of your company's contract?	☐	☐
If not, have you planned an efficient review process?	☐	☐
Is the negotiation calendar in place to meet the expected or required timing to close?	☐	☐
Is your calendar in place to manage the negotiation flow?	☐	☐
Will your team members be available as required?	☐	☐
Is the process in place to keep team members informed?	☐	☐
Have you reviewed with your team how this negotiation process will proceed?	☐	☐

Now apply what you have learned in this chapter to the following situation.

You are negotiating a deal with a company we'll call "IOT Devices". This is a big deal for both sides. This transaction will give them product to sell for half of their $100-million plus annual revenues. The two teams are five months into negotiation, which has gone smoothly. As you and *your team work out the deal, the IOT Devices developers keep calling your technical people with questions.*

Your technical team member realizes what is happening, she says, "IOT Devices has ordered systems and is finishing porting their applications to your platform, but we haven't yet finished the deal." You and your team see the position that IOT Devices have put themselves in. They need you to make this deal more than you need them. The deal comes down to three elements of pricing, as follows:

1. Services (40%)

2. Software (30%)

3. Hardware (30%)

Today's negotiation opens when IOT Devices' lead negotiator, Rick, leans over the table and says, "Let's talk about pricing on services." Okay, you think, we'll talk about pricing because it is the only key term left. Then Rick says, "You guys are way out of line with prices on your services! W-a-y out of line! We won't let you get away with such highway robbery. You need to speak to our V.P. of Services."

You and your team members are puzzled because the negotiations with IOT Devices over the past months have never been like this. One team member questions whether they are playing pricing games. Why is Rick being so adversarial—especially when he needs this deal badly? Nonetheless, you strike a deal with the services V.P. on prices. You give them a little lower price than you initially expected, but still stay within your NSR.

Now comes the software negotiation. You and your team have caucused and decided on a strategy: Offer a higher price than you had planned. You do this because you sense they will be playing the "escalation game." You want a cushion in place in case you must make pricing concessions. Rick once again responds angrily, "My software V.P. says you're way out of line." The software V.P. calls to express outrage at your starting price and you let him vent.

Finally, after looking at your financials, you lower the price in a principled manner. However, since you and your team are now aware of IOT Devices' strategy, the new price is still a little higher than you would have otherwise given them. Now, you have broken the code on their tactic. They are reviewing the initial offer and, with aggressive indignation, escalating to get a lower price.

We handled the hardware discussion much like the software one. We held back on price, giving them a higher price than we would have in the absence of the escalation tactic. After reaching agreement in the hardware pricing discussion, we received a call from the IOT Devices lead negotiator. Rick said, "My senior VP still thinks the overall price is too high. He wants to meet with your line of business President. We will fly to your headquarters."

STOP if you are using the Companion Workbook.

Exercise 15-4: Time and teamwork

Prior to the meeting, we met with our line of business President and the team to discuss strategy. Internally we agreed that we would give them an additional 1.5% incentive if they signed the deal that day. We set the following agenda for the meeting with their senior VP:

09:00-09:30	Teams meet. IOT Devices presents concerns.
09:30-10:15	Our team meets internally to review IOT Devices' requests.
10:15-10:30	Our client's V.P. explains to IOT Devices that we have to do more analysis. (Why? Because of Principle #4: "Concessions easily given appear of little value.") We also don't want to have other issues appear before they have to leave.
10:30-Noon	Caucus.
Noon-13:30	Teams join for lunch.

13:30-16:00	Discuss implementation of the agreement.
16:00	Our client's V.P. returns to offer the additional 1.5% incentive if they sign today and provide a reference.
16:30	IOT Devices leaves for the airport.

Did it work? You bet it did! At 4 p.m. (16:00), our client's V.P. said to the IOT Devices team, "It cost you about $8,000 to come in for the day. I have a great return for the $8,000 you invested. We scrubbed the financials again, and if you sign now and become a reference for us at least four times over the next year, I will give you a credit for 1.5% which equates to $450,000." And so, they did.

The K&R Deal Forensic

Our team did several things right with this deal. The customer's escalation tactic was quickly recognized. The team had very good communication and recognized its leverage. Leverage was used wisely. Face was given to the other side. Concessions made were principled. Teamwork was great at all levels.

Our client ended up making 40% gross margins—a very high margin for this type of deal. The teamwork involved us working with our client's day-to-day negotiation team and with executive management. This situation shows the importance of teamwork on all levels. Not only did this deal work for both sides, IOT Devices felt great, too. Their senior V.P. got a great return for his investment of time for the trip and went home a hero. (Remember: "Face".)

Teamwork = best minds and best chance for success.

WHY DO SOME DEALS SUCCEED...BUT OTHERS FLOP?

Deals succeed because of teamwork, patience, and listening. You must be prepared and carefully weigh (and create!) leverage. Your skill and patience will determine how well you use your leverage. Your skill will also determine how well you will do when the other side has more leverage than you do.

We've covered a lot, so take a few minutes to review. In the workbook, write the top twelve reasons why you think deals succeed.

STOP if you are using the Companion Workbook.

Exercise 15-5: Success is ours!

Here are the **K&R Top Twelve Attributes of Successful Deals** (that lead to lasting relationships):

1. Your true goals are defined, understood, and agreed upon.

2. Value is articulated (and quantified) with persuasive rationale.

3. Your weaknesses are managed; theirs are utilized.

4. Leverage is understood, maintained, and used wisely.

5. Strategic and negotiable parameters are understood.

6. Attitude is positive, not adversarial.

7. The right resource is engaged.

8. Executives buy in on goals, timing, and strategy.

9. Mandatory goals are achieved through acceptable solutions.

10. Both sides perceive the deal is within the NSR.

11. Governance and responsibilities are defined and respected (including regular communications).

12. The deal is deliverable; implementable.

Bonus: **You need to be able to implement your contracts**. Do not put obligations in a contract that you cannot meet. It can impact credibility, lead to disputes, damage relationships, and even become a deal breaker.

Do not underestimate the other side. Assume everyone you encounter in negotiations is intelligent, and assume deals will only succeed if they work for both sides. Each side may have different goals in a deal. Goals can be profitability or revenue, but they can also be cost reduction, asset transfer, market acceleration, full employment, political recognition, and so on.

Everyone wants to win, but an equal win-win is rarely the outcome. Why? Because winning is uneven. That's because goals are different and leverage is usually unequal. The measurement of success is different between the parties. Therefore, there is the potential that one party will do better than the other, or that both sides will feel they have won more.

You always want to give "face" to the other side, even if you feel you have gotten the big win. The deal needs to work for the other side or you will be back at the negotiation table sooner than you think...or even worse, the relationship will end prematurely at a high cost.

Chapter 15: The Importance of Teamwork

WHAT YOU LEARNED IN THIS CHAPTER

- Create a balanced team that can best deal with the issues on the table.

- Keep your negotiation team united.

- Debate internally; unify externally.

- Recognize varied motivations, both internal and external.

- Build two-way trust and responsibility with management.

- Create a true team in which everyone contributes expertise.

- Work as a team: Communicate openly within the team.

- A lead negotiator who has earned the confidence of team members has a much better chance of minimizing interruptions; team members will trust that the lead negotiator will get to the point even though the path may be different from what they might have taken.

- Teamwork report card: If your deals create lasting relationships, you deserve a high grade.

- Remember our twelve top reasons why deals succeed (and lead to lasting relationships).

CHAPTER 16: CONGRATULATIONS!

We hope that you enjoyed the participatory nature of this book. It was designed to get you to think as an active negotiator. We haven't let you relax much, but we know that you are excited about negotiating. We covered how you can grow your business results, and improve your bottom line as you negotiate with the confidence that comes from knowledge, credibility, and creating value, impact and outcome. You've seen that successful negotiations come from preparation and planning, as summarized by our Six Principles:

K&R'S SIX PRINCIPLES™

1. Get M.O.R.E.—Preparation is key to a winning negotiation.

2. Protect your weaknesses; utilize theirs.

3. A team divided is a costly team.

4. Concessions easily given appear of little value.

5. Negotiation is a continuous process.

6. Terms cost money; someone pays the bill.

But before we let you go, let's take a moment to discuss K&R's Big Three Deal Makers or Deal Breakers that pull together much of what we have discussed.

What are the "Big Three" factors? They are:

- Price

- Product/service and their value (business impact)

- Ownership/control

If two sides come to an agreement on price, product/service, and ownership/control, the deal will usually get done. From the early stages of the negotiation process onward, these factors deserve your attention. Let's look at each factor in detail.

PRICE

Price refers to the basic price term and its relationship to the value received. There are many price-related terms that have varying degrees of importance to either side (remember MID), and can swing the value of the deal in favor of you or your counterparts. These include:

- Timing of payments

- Late payments

- Cost of service and support

- Delivery costs

The following example from our files illustrates a creative way of making a good deal and establishing a relationship when price is an issue.

> *A large customer we'll call "Chapter 11" was financially shaky.*
>
> *They wanted additional discounts on future product shipments.*
>
> *"Chapter 11" had considerable accounts receivable from major Fortune 500 companies, which were pledged as securities to their creditors. Instead of lowering the price or extending additional credit, we negotiated with the creditors to allow some of those receivables to be assigned to us at a discount. That way, we were able to make our sale at our prices and get compensated for time, value, money, and risk. The customer got the products they needed to generate revenue, and the creditors helped the company remain viable to secure their interest for the future.*

Chapter 16: Congratulations

In essence, we were acting as a finance company. We were creative in solving their financial problems.

PRODUCT/SERVICE

There are many product/service related terms that influence the overall offering. These include:

- Scope of service
- Scope of license
- Technology differentiators in a resale environment
- Statement of work for development
- Support of products
- Personnel decisions on services
- Roles and responsibilities of both sides

And don't forget that one of the things that most influences the overall offering is the value you and your solution brings to them; in the end, value is the measurable impact of delivering their outcome.

But here's the bottom line: If the parties agree that the product or service being described is appropriate for achievement of their goals, the deal can be done. In a joint development context, this means agreeing on the end result of the development and the elements to be contributed by each side.

OWNERSHIP/CONTROL

This can be the most difficult set of issues to resolve, particularly in a services joint development or custom development deal. The parties must seek agreement on intellectual property ownership and rights. In joint development, it also means control over the development process, change management, engineering change rights, and so on. When it comes to intellectual property, particularly in the joint development scenario, it is extremely important to understand who has what rights regardless of whether the venture is successful or unsuccessful.

In services engagements, especially when they are long-term, negotiation over a clear governance process that addresses these issues is important to ensure smooth delivery, especially as personnel change and memories fade.

Use your leverage wisely to earn trust, build value, and negotiate on the merits of the transaction. Create value in a deal from advantage given to both sides.

THANK YOU AND AU REVOIR (GOODBYE FOR NOW)

We want to reiterate that there are many exceptions to the general principles discussed in this book. The key is to understand those principles and do your homework in preparation for the deal. Then you will make _conscious decisions_ when making an exception to the general principles of the K&R Negotiation Method™. Those types of conscious decisions lead to prudent risks that we often take in today's business environment. Because the risks are consciously taken, you will be able to deal with the usually predictable consequences when plans go awry.

In closing, we want to remind you that negotiations will always involve balancing risks and rewards. So, stay focused on the merits of the transactions and you will do just fine. Above all, enjoy the experience!

Mladen Kresic

and your K&R Negotiations team

GLOSSARY

communication: the process of creating meaning.

compelling argument: persuasive argument which compels a party to take action because it expresses value to them.

credibility: the ability to inspire belief. Credibility is the cornerstone of persuasive communication.

credible offer: an offer based on sound business rationale. A credible offer is believable because it has rational underpinnings. It becomes persuasive when the rationale is related to value to the other side.

effective persuasive communication: communication that is understood, believed (or credible) and contains a statement of value for the other side.

gamesmanship: techniques or actions unrelated to the merits of the transaction that are used to gain an advantage in a negotiation. Gamesmanship is a subset of tactics. For example, yelling, screaming, or walking out. See also - tactics.

goals: the ends, your purpose for negotiating (should not be confused with the means of getting there).

K&R Negotiation Method™: proven process and tools for preparation, planning, and execution of negotiations described in this book, a part of K&R's Win-Wisely Methodology.

leverage: in a negotiation context, leverage is the ability to persuade people to move closer to your position. Leverage is both fact-based (I have unique value to you) and skill-based (I know how to articulate my unique value to you).

linkage of terms: the interrelationship between agreement terms. The most obvious linkages are between performance and enforcement terms. For example, "Payment" is a performance term. Penalties for late payment, termination rights, and audit rights are all enforcement terms.

listening: hearing something with thoughtful attention.

macro agenda: the entire negotiation schedule (or timetable) of activities and application of the right resources from beginning of discussions through deal closing, including internal discussions, timing of document creation and exchange, and meetings.

means: alternate ways of accomplishing goals.

micro agenda: the plan and

execution of a single meeting or interaction internally or externally (could be by telephone, for example) which includes pre-meeting preparation.

MID™: A tool within the K&R Negotiation Method™ for distinguishing among mandatory, important, and desirable requests in a negotiation. Also used to identify conflicting requests and to distinguish between means and ends.

momentum to closure: the inertia that drives people to close which occurs once all major issues in a negotiation have been resolved. Closing may occur even though not all of the minor issues have been addressed. Instead, these issues default to the way they are in the document draft.

negotiation: discussion between or among people to reach agreement.

Negotiation Capital©: the amount of willingness by either side to negotiate and exhibit flexibility with the other. For example, my Negotiation Capital is your willingness to be flexible in negotiations with me. Negotiation Capital is a function of value you provide and trust you build with the other side. This capital can be "spent" as leverage to move the other side closer to your position. Or it can be "burned" by making mistakes or losing credibility.

Negotiation Success Range (NSR)™: the boundaries within which both parties will be satisfied with prices and terms.

patience: taking the time to work through the process of negotiation.

persistence: making repetitive arguments; the "offensive" or proactive form of patience.

principled concession: concession made with credible business rationale usually related to value.

psychological stake: a personal, often emotional motivation that replaces or impacts logic to get the deal done. It is also the tactic of creation or exploitation of people's psychological stake.

quantified value: impact that is understood and measurable.

quantify: make understandable in measurable terms.

successful negotiation: reaching agreement on terms that satisfy your goals and theirs to varying degrees.

tactics: techniques or actions intended to influence a negotiation. For example, great teamwork and coordination to get information useful to your negotiation are types of tactics.

value: what the proposal that you make to the other side does for them. In a sales context, value is the relationship between price and function that should be related in a quantified way to a business impact that the buyer wants or needs to achieve.

value argument: a persuasive argument because it expresses benefits that help reach a goal the other side wants to achieve.

NEGOTIATE WISELY IN BUSINESS &
TECHNOLOGY

www.ingramcontent.com/pod-product-compliance
Lightning Source LLC
Chambersburg PA
CBHW081457200326
41518CB00015B/2291